The Privatization of the Oceans

The Privatization of the Oceans

Rögnvaldur Hannesson

The MIT Press
Cambridge, Massachusetts
London, England

First MIT Press paperback edition, 2006
© 2004 Massachusetts Institute of Technology

MIT Press books may be purchased at special quantity discounts for business or sales
promotional use. For information, please e-mail special_sales@mitpress.mit.edu or
write to Special Sales Department, The MIT Press, 55 Hayward Street, Cambridge,
MA 02142.

Set in Stone sans and Stone serif by SNP Best-set Typesetter Ltd., Hong Kong.
Printed and bound in the United States of America.

Library of Congress Cataloging-in-Publication Data

Hannesson, Rögnvaldur.
The privatization of the oceans / Rögnvaldur Hannesson.
p. cm.
Includes bibliographical references and index.
ISBN 0-262-08334-5 (alk. paper), 0-262-58265-1 (pb).
1. Fishery management—Economic aspects. 2. Fisheries—Limited entry
licenses. I. Title.
SH334.H362 2004
338.3'727—dc22   2004044963

10 9 8 7 6 5 4 3 2

Contents

Acknowledgments   vii

Introduction   1

1   Economic Welfare and the Evolution of Property Rights   7

2   The International Law of the Sea   29

3   Property Rights in Fisheries   43

4   Toward Individual Use Rights   69

5   Successes and Failures: New Zealand, Chile, Norway, and Canada   85

6   ITQs in Iceland: A Controversial Reform   113

7   The Development of ITQs in the United States   135

Conclusion   163

Notes   179
References   191
Index   199

Acknowledgments

I started working on this book during a sabbatical stay at the University of Washington's School of Marine Affairs in the first half of 2002. The working environment in Seattle proved excellent. The faculty of the School of Marine Affairs is an invaluable source of knowledge not just about the fisheries of the United States but also about fisheries in many other parts of the world, and the University of Washington has an excellent library. I am grateful to the School of Marine Affairs for hosting me, and to the Norwegian School of Economics and its Institute for Research in Economics for granting me a leave of absence and providing financial support.

As the manuscript evolved, several friends and colleagues read portions of it, corrected factual misstatements, and gave good advice. Dave Fluharty and Dan Huppert, both at the School of Marine Affairs, read the chapter on the United States; I hasten to add that our views on the fisheries management councils diverge. Ásgeir Danielsson, now at the Central Bank of Iceland, and Ragnar Árnason, at the University of Iceland, read the chapter on Iceland; Ásgeir provided much help with data on quota prices in Iceland. Julio Peña-Torres, at the Alberto Hurtado University in Santiago, checked the material on Chile and provided much information on the development of the fisheries management regime in that country. My colleague and friend of old, Lee Anderson, read the whole manuscript and provided many good comments. I shall exonerate all these gentlemen from any responsibility for the final product, and they will indeed find that I did not follow their advice on a number of occasions. Sandra Valenzuela, a librarian at the Biblioteca Nacional, responded patiently to a series of enquiries about a map supposedly published in a Chilean magazine in 1940.

At various places in the book I discuss issues that I dealt with earlier in the proceedings of a conference held by the Center for Free Market Environmentalism (PERC) in 2002 and in documents prepared for the World Bank ahead of the World Summit on Sustainable Development, held in Johannesburg the same year.

The Privatization of the Oceans

Introduction

This is a book about economic institutions. More precisely, it is a book about economic institutions and their development in world fisheries. Why the fisheries? In most countries the fisheries are a very small part of the economy, although important in some countries and regions as a source of nutrition and even national wealth. But the fishing industry illustrates the emergence and role of economic institutions better than many other, more important industries and is therefore worthy of attention far out of proportion to its share of world production. There are several reasons for this.

First, the oceans are, or were, the last commons. No single state used to have jurisdiction at sea outside a narrow belt, which as late as the middle of the twentieth century was only 3 nautical miles wide. Without a wider national jurisdiction at sea, it is hard to imagine how an economic institution such as property rights could have developed for any but the most stationary fish stocks. People who still have not reached the age of retirement have in their lifetime witnessed a revolution in the international law of the sea, by which states have gained control over fish resources off their shores. In the wake of this we have seen exclusive individual rights of access to fish resources develop. What we have before us is a development of economic institutions which in other sectors of the economy is partly or wholly shrouded in myth.

Second, the fisheries illustrate better than many other industries the consequences of weak or absent property rights. Fish accessible to anyone will be captured as quickly as possible by those who first discover them, for otherwise they will be taken by somebody else. But while rational for each individual fisherman, this can be a counterproductive and outright dangerous practice for all fishermen taken together. Fish grow with age, and fishermen would be better off taking fewer but larger fish rather than many and small. Furthermore, some fish must survive to reproductive age and therefore should not be caught prematurely. But without rights securing for individuals the fruits of their own restraint, no one will have an incentive to leave anything behind. That fish stocks have not become extinct or nearly so like the great auk and the buffalo owes more to nature hiding them in an impenetrable body of water than to any restraint by humans.[1]

Third, the battles that have taken place in countries that have tried to put in place property rights in their fisheries are extremely interesting illustrations of the forces that cause and mold changes in economic institutions. The driving force is a perception of gains, either by industry players looking after their own interests or by politicians and civil servants who see property rights as beneficial for society, and usually both. But even if a new institution such as property rights to fish would bring an overall gain to society it does not necessary benefit all and harm no one. While those who expect to gain will promote and support the new institution, those who expect to lose will fight it with equal or greater vigor. Sometimes the gainers prevail, but at other times the losers do.

More is involved, however, in these battles than a fight between clearly defined gainers and losers. The division into "gainers" and "losers" is unduly crude. Some gain more than others, and some lose less than others, and who gains and who loses is both subject to uncertainty and dependent on how exactly the new arrangement is designed. The political authorities that finally decide whether or not to put in place property rights in fisheries, and how these rights are to be defined, are therefore subject to a multitude of forces pulling in different directions, and there are many possibilities with regard to what the final outcome will be. The outcome is more often than not a compromise deviating from the blueprint an economic-institutional analysis might come up with. Sometimes the "compromise" might be designed so as to make the regime unworkable, as seems to have happened with the so-called Merino Law in Chile, to be discussed in chapter 5.

The role of property rights in fisheries is no different from the role of property rights elsewhere in the economy: property rights, if adequately defined and enforced, encourage efficient use of resources in the present with an appropriate regard for the future. With the socialist experiments now all but abandoned, there seems to be no serious alternative to an economic system based on property rights. Yet the establishment of property rights to fish has met with surprisingly strong resistance, even in countries that otherwise seem strongly committed to property rights and market capitalism. Perhaps that system would not have been put in place if we had started from scratch; there is a case for saying that its virtues are somewhat, perhaps even strongly, counterintuitive and need to overcome other, unappealing features. However that may be, the fisheries are but the last of the common property resources to which private property rights have developed; recorded history tells of enclosures and clearances of common land in response to changed economic circumstances. This process has many parallels to what has happened in fisheries: increased scarcity of resources, gains for those who could claim these resources for themselves, the impossibility of accommodating all those who wanted to stake a claim to the resources without a major loss in productivity, bitter fights over rights to access, and the shifting fortunes of those who are excluded from enjoying the free access they once had to land or other resources. Chapter 1 sets the stage by revisiting the famous enclosures in England and the clearances of the Scottish Highlands.

The enclosure of the world's fish resources began as an attempt by states with rich fisheries off their shores to extend their jurisdiction over these areas and to clear away foreign fishing fleets. This development was enormously stimulated by the claims to exclusive national rights to offshore oil and ended in the establishment of the so-called exclusive economic zone. Without this jurisdictional framework it would not be possible to limit fishing except by agreement among an indefinite number of states, an outcome that is none too likely. Chapter 2 traces the development of the law of the sea since the beginning of the sixteenth century. It is a fascinating story, and the development has not been straight in one direction. From the early 1500s to the 1800s the world went from wild claims to exclusive rights on the sea to a general acceptance of freedom of navigation and fishing. The growing scarcity and value of resources such as fish and offshore oil made that kind of free-for-all increasingly anachronistic and wasteful.

While in principle property rights to fish could be the solution to a number of problems, such rights are difficult to define and enforce in practice. Fish do not respect any boundary lines on charts, and they cannot be branded and individually monitored by their presumptive owners.

Therefore, ownership rights to fish swimming freely in the sea are conspicuous by their absence. What can be established, however, are use rights such as rights to catch a certain quantity of fish or to apply a certain effort to catching fish. Chapter 3 discusses the role of property rights in fisheries and what form these rights may take.

Exclusive use rights to fish have been established in a number of places, and we seem to be in the middle of a process where this is happening worldwide. Chapter 4 discusses the main driving forces behind this development. Sometimes it has been individuals acting collectively to promote their interests, but also, and perhaps more often, the initiative has come from government circles in an attempt to increase efficiency overall as well as in the fishing industry itself. Much controversy has revolved around whether these novel use rights amount to private property, like the ownership of land. The question of formal property rights is not unimportant, but perhaps it is less important than many people think. Private property rights are usually subject to constraints, at the very least the constraints embedded in the law of the land and in customs of conduct.

One particular form of exclusive use rights in fishing is fish quotas, held by individuals or firms, which can be leased for a limited period of time or bought and sold permanently. These are usually referred to as *individual transferable quotas* (ITQs) or *individual fish quotas* (IFQs). There is reason to believe that such use rights are particularly useful to promote efficiency in the use of fish resources. Three chapters are devoted to the development of ITQs in various places around the world, what they have accomplished, and what has hindered versus promoted this development. This treatment does not cover all cases worldwide; some readers may, for example, miss the quota system in the Netherlands, which was one of the first to be put in place. I have chosen cases which I felt I knew reasonably well and which illustrate how different processes and outcomes can be, not just between countries that appear similar in terms of economic institutions and development (Iceland and Norway, for example), but even within a country (the United States).

Chapter 5 deals with developments in New Zealand, Chile, Norway, and Canada. New Zealand and Canada are cases of success; Chile and Norway are cases of failure. New Zealand has gone further than any other country in establishing private use rights to fish and is also among those who were early in the game. In Chile a fresh law establishing ITQs was radically changed after a democratic government came to power in 1990. An attempt to establish ITQs failed in Norway in 1991. Norway has, however, long had a system, or systems, of exclusive fishing rights through which fish quota allocations have been indirectly tradable. Development is under way both in Chile and in Norway to move closer to marketable fishing rights, in Chile in the form of ITQs and in Norway through strengthening the tie between fishing concessions and tradable quota rights of fishing vessels. Canada does not have a national policy of promoting ITQs but has allowed them to emerge where the industry has been in favor. The Canadian examples are quite interesting because they in some ways contradict received wisdom about reforms in fisheries management.

Chapter 6 discusses the case of Iceland, which, along with New Zealand, introduced ITQs early. It started as a response to an emergency that people hoped would go away, namely a severe depletion of the cod stock, the most important fish resource of the country. But the quotas were there to stay, and even if there was partial backtracking for a time the quotas developed into a permanent system, or as permanent as any system which can be changed through legislation. The smallness and the extreme fishery depen-dence of Iceland give a greater urgency and prominence to questions regarding fisheries management than is the case in most other places. The Icelandic case is also interesting because it has given a sharp focus to the question of efficiency versus equity; few people doubt that the ITQ system has increased the efficiency in the Icelandic fisheries, but many find the distribution of the associated benefits lopsided.

The United States has also experimented with ITQs. The United States is, as everyone knows, a vast country where local conditions vary greatly. It is therefore not surprising that the responses to problems in the fisheries of the United States have been different in different places. To some extent those differences are dictated by differences in circumstances, although that is not the whole story; ITQs seem about as pertinent a solution to the problems of New England as they have been in Alaska and the Pacific Northwest. Chapter 8 deals with four cases of ITQs in the United States: the surf clam and ocean quahog fishery off the Atlantic coast, the Alaska halibut fishery, the fishing cooperatives in the whiting fishery in the Pacific Northwest and the Alaska pollock fishery, and the emerging "three-pie" quota system in the Alaska crab fisheries. The idiosyncrasies of the American political system have led to an innovative approach to establishing private fishing rights; while ITQs were for a period banned up front, they were introduced through the back door through so-called fishing cooperatives by a special legislative action known as the American Fisheries Act.

The development of the ITQ systems is a fascinating story that varies from place to place. Sometimes it has not been successful. The final chapter tries to draw some conclusions and see common traits. The development of ITQs involves much fighting among interest groups; there is strategic positioning among those who look set to gain, and there is opposition from those who fear being left out. Few changes are such that no one loses, even from changes that can be highly beneficial in the aggregate. How much obstruction from potential losers is legitimate? Are all losses legitimate objects of concern, even those that involve elimination of inefficiencies associated with a poor system of management? Any regime, even a bad one, creates its own vested interests. Would the ostensibly reasonable requirement that all losers be compensated in fact obstruct progress and ultimately be inequitable?

Establishing property rights where there were none before involves legislation and case law. Ideology and the general outlook among judges and legislators and among the general public affect the evolution of legislation and court verdicts. This process is complex and can change direction more than once under contradictory influences. The development of economic institutions is, therefore, an evolutionary process. The final outcome, if there can be such a thing, seldom corresponds to an ideal blueprint for solving specific problems; there are too many competing interests affecting the process for that to happen, too many designers acting independently. But what survives is what works, what serves a purpose.

1 Economic Welfare and the Evolution of Property Rights

The economic system in what we usually call Western industrialized countries is based on private property. Property rights extend not just to small items for personal use or larger items for one's immediate family such as dwellings; they also extend to means of production, and land and its various resources. It can be argued that property rights to means of production and land is a fundamental reason for the success of this economic system. This arrangement also is, or has been, highly controversial. Socialism arose in protest against it and its perceived injustice. That perception was not without foundation. The Industrial Revolution made those who owned land and means of production immensely rich, but it is doubtful whether it made others poorer, at any rate in absolute terms. Pre-industrial societies were anything but affluent and egalitarian. Some individuals have always been able to get an edge over others, either by birth or by knowing how to maneuver themselves into positions of power and privilege. The Industrial Revolution and the economic development that followed raised the living standards of all nations in which it took place and brought "the common man" riches that his ancestors eking out a precarious subsistence would never even have dreamed of.

Yet the societies undergoing the Industrial Revolution or having just emerged from it were marred by a skewed distribution of wealth and the class struggle which it generated. Less than 100 years ago, socialism was a still untested dream of a more just and harmonious society. Many believed in it and offered their lives for it. As the practical experiments with socialism got under way, first in the Soviet Union and later in other countries, the dream became harder to believe. It finally ended in a nightmare, and the experiment collapsed when the Soviet Union, the first socialist state, fell apart. Most of the few remaining states with a nominal association with socialism are now busy distancing themselves from its economic principles and practices.

What brought the experiment down? Could it have ended otherwise? The diehards would say that socialism has not really been tried yet. Even if the downfall of the Soviet Union has many and complex causes, one seems most important: the socialist system simply was not productive enough. The material standard of living in the affluent "West" further and further outpaced that in the socialist "East." Even if the socialist East did its best to keep out Western media, awareness of the difference trickled down to the general public in the socialist countries, and members of their elite who traveled abroad became painfully aware of it.

This was not always so evident. In the 1950s and the 1960s, when Sputnik went aloft and the Soviet Union demonstrated the quality of its science and the prowess of its technology to the entire world, many people in the capitalist West, economists among others, believed that the socialist economies might one day overtake the capitalist economies in terms of material production. The economic growth of the Soviet Union seemed impressive; the fact that it still lagged behind the United States could be explained by a lower starting level and the devastation of World War II. The famous American economist Paul Samuelson published a diagram in the 1961 edition of his classic introductory textbook showing that the Soviet Union might overtake the United States in gross national product by the year 2000. The text accompanying the diagram reveals the undecided nature of the economic contest between the two superpowers at the time. "All seem to agree that [the Soviet Union's] recent growth rates have been considerably greater than ours as a percentage per year. . . . It will be evident that the Soviet Union is unlikely to overtake our real GNP for a long time to come, and our per capita welfare level for a still longer time to come. . . . [O]ur two systems are on trial in the eyes of many uncommitted underdeveloped nations." But Samuelson concluded "on a note of optimism. . . . Our mixed economy—wars aside—has a great future before it. Writing a textbook some 30 years ago, one could not have said all this: looking around at the shrinking international trade network, at the collapsing banking structure, at the grim specter of poverty midst plenty, some might then have despaired over the future of free societies."[1] This last quotation may serve to remind us that, however productive capitalism may be, it will survive in democratic societies only if it succeeds in distributing its fruits reasonably equitably.

Private property rights are not the whole story behind the success of Western capitalism; the issue is immensely more complicated than that. Pre-industrial societies did not lack private property rights; they lacked a technology and an organizational framework that would have made it possible to use such rights productively. What capitalism and the Industrial Revolution accomplished was to mobilize the surplus value produced by labor (i.e., the value over and above what was needed to maintain and reproduce the labor force) for investment, making a still greater surplus possible. Many societies of the past were rich and produced substantial surplus value, but that value was appropriated by a predatory and unproductive ruling class and by the church or other religious or ceremonial institutions. The cathedrals of Europe, the pyramids of Egypt, and the temples of Thailand are among the legacies of this past. And societies did not have to be very productive to do this; the Easter Islanders converted their meager surplus production to carving statues out of their mountains and transporting them over long distances to the places where they were erected, their stony faces staring sternly at the lowly inhabitants who lived in primitive huts they could not even enter upright. How the limited technological knowledge of the Easter Islanders enabled them to do this still boggles the minds of those who try to understand it.

That private property should be among the keys to general prosperity is more than a little paradoxical. Private property is a manifestation of self-interest and greed; if we were happy to share everything, there would be no reason for private property. It is indeed paradoxical that a system based on self-interest and greed has proven itself superior to socialism, which is based on shared interests and common ownership. Dreamers of all ages have found it difficult to come to terms with this. No one has, perhaps, expressed it more eloquently than the famous French romantic Jean-Jacques Rousseau[2]:

The first man, having enclosed a piece of land, [who] thought of saying "this is mine" . . . was the true founder of civil society. How many crimes, wars, murders; how much misery and horror the human race could have been spared if someone had pulled up the stakes and filled the ditch and cried out to his fellow men: "Beware of listening to this impostor. You are lost if you forget that the fruits of the earth belong to no one!"

And no one has replied more eloquently than Rousseau's countryman Voltaire, who scribbled this in the margin[3]:

What? He who has planted, sown, and enclosed some land has no rights to the fruits of his efforts? Is this unjust man, this thief to be the benefactor of the human race? Behold the philosophy of the beggar who would like the rich to be robbed by the poor!

And in a letter to Rousseau, Voltaire added this[4]:

I have received, monsieur, your new book against the human race, and I thank you. No one has employed so much intelligence to turn us men into beasts. One starts wanting to walk on all fours after reading your book. However, in more than sixty years I have lost the habit.

Private property and self-interest constitute a powerful incentive mechanism. Being assured of the fruits of his efforts, the owner of a piece of land, a factory, or a mineral deposit has an obvious interest in taking good care of it and using it in the most productive way. Furthermore, a system of ordered and accepted property rights avoids devastating struggles over what would otherwise come into and remain in one's possession through taking and defending by force. Finally, as was emphasized by Hernando de Soto, secure property rights make it possible to "mobilize" property by using it as a collateral for credit to initiate new, productive projects or expand existing ones.[5]

Would humanity have chosen an economic system based on self-interest as a driving force and private property as a mode of organization if we were to design it from scratch, not knowing what would work and what would not? Probably not. Chances are that we would find it repugnant, and that we would instead go for a system with a more sympathetic appeal, one based on common property and care for our fellow human beings (this is what socialism was supposed to be about). But market capitalism was not designed from scratch, and it did not descend upon us all of a sudden. Market capitalism has evolved over a long period of time, through small changes and adaptations of institutions, and historians can probably argue endlessly about what got it going. It is, as the Scottish philosopher Adam Ferguson put it about jurisprudence, a "result of human action and not of human design."[6]

Another reason why the victory march of market capitalism is somewhat surprising is that this system seems to be a recipe for chaos rather than coordination. In capitalist market economies there is no single coordinating institution; decisions are made by individuals, on the basis of their (or their employer's) self-interest. It is not obvious how all these decisions interweave into a coherent whole capable of satisfying human needs in an acceptable manner. As Thomas Schelling has aptly noted, there are many cases where the overall consequences of individual decisions are unexpected and maybe unacceptable, yet unintended.[7] The sum of individually innocuous parts can be evil. Worries about the coordination failures of market capitalism have compelled many economists to propose that the state should govern the economy in some detail. Indeed, there have been times, especially during the Great Depression, when capitalist market economies did not seem to be functioning well, as the above quotation from Samuelson alludes to. Today, however, few ideas have lost currency to a degree comparable to the "economics of planning."

That social welfare can be maximized through individual pursuit of private interest has been known since Adam Smith if not longer. Smith's famous phrase "as if guided by an invisible hand" does little, however, to explain how the coordination problem is solved, and it must still be regarded as a bit of a mystery how in fact this is achieved in a market economy. To some extent coordination may be fortuitous and specific in time and space; coordination failures such as the Great Depression have in fact happened, and rags and riches continue to coexist, more so in some places than in others. The advantage of the market economy lies perhaps first and foremost in making use of information where it is available, as emphasized by Hayek, and its release of individual energy through its appeal to individual gain. The seeking of self-interest is, however, like a powerful beast. If it runs amok, it may destroy; if it is tamed and harnessed, it will do useful work. Market capitalism works wonders when it is well tamed and harnessed. Unfettered capitalism is not a pretty sight. The Russia that emerged from the wreckage of the Soviet Union is a warning example.

So, in order to fulfill its role as a useful system of organizing economic activity, market capitalism has to be supplemented with a governance structure that channels its energies for the common benefit. The productivity of the system is one aspect, the distribution of its results another. All market capitalist societies are characterized by unequal distribution of wealth, but as long as the system is perceived as delivering the goods in a reasonably equitable manner this can be tolerated, and even welcomed if it is seen as a precondition for productivity. At least in democratic societies, an economic order that is perceived as grossly unfair is not likely to last long, and in authoritarian societies elites resting on bayonets always lead an uncertain existence.

Despite the usefulness of and the powerful incentives associated with private property rights and market transactions, there are limits to how far they can and should extend. It is possible to come up with economic arguments in favor of slavery; the slave owner would have a stronger incentive than an employer of free labor to provide his slave with skills because the slave could not voluntarily change masters, but few of us would think that the argument stops there. Modern medicine has created a basis for a market in organs, but does that mean that people should have a right to sell themselves (or their offspring) in parts? Many people today are desperate enough to find that an attractive proposition.[8] But in between clear-cut cases there are many which are less so; there is an element of judgment in how property rights should be defined and circumscribed and what should be left to markets to sort out and what should not. The subject of this book provides a fairly clear-cut case: as long as we regard fish primarily as a source of food and other material benefits, the problem in fisheries worldwide is absence of property rights and market transactions rather than the opposite.

Origins of Property Rights

How did private property arise? Did someone, as Rousseau put it, put down stakes or dig a ditch and say "This is mine" while others just said "Lo and behold"? Hardly. Private property is the outcome of the stronger arrogating to themselves a piece of land or whatever and keeping what they have built or made or occupied by being ready and able to defend it. This was a collective effort. In primitive societies one tribe defended itself against another tribe. Later the lord and his men defended themselves against other lords and their followers. Leadership, undoubtedly, has always been important; some have always been "more equal than others." Sometimes leadership institutions become fossilized. There was a time when kings led their forces into battle, a time when being a king was a risky occupation.[9] In those modern democratic societies that have not dispensed with them altogether, kings are ceremonial institutions.

Private property rights to land probably developed as a practical institution. The tribal chieftain rewarded his most trusted men with rights to certain pieces of land. The strengths of these rights have evolved over time; feudal lords fought with the crown for stronger rights, tenants with the lords. In medieval villages in England and much of Europe, much land was held in common, but individual families had private or semi-private plots scattered around, sometimes on a rotational basis. They had obligations to render services to the lord, sometimes converted into payments, whereof tenancies.

Even if much of the development of property rights is lost in the mist of unrecorded history, we do have written accounts of how property rights have been established in new, uninhabited lands or in lands sparsely settled by what in earlier times would have been called primitive tribes. When the Norwegian settlers came to the uninhabited island they named Iceland, rules were developed with regard to how much land each individual could claim. Men could take as much land as they could enclose by bonfires in one day, women as much as they could tow a cow around over the same period.[10] These rules applied to the leaders (and even in those days, women could be leaders); each leader had in his or her household a band of laborers and slaves. Needless to say, the latter got no land except at the discretion of the leader. Gradually, and undoubtedly for practical reasons, some of them got their own plots, and the slavery disappeared. In the United States, groups of settlers often appear to have come to agreement among themselves about rights to land and to minerals. Sometimes these "rights" were to land claimed by other settlers of European origin and hence tenuous. Hernando de Soto tells of how such grassroots and squatter rights came to be recognized and incorporated into the law of the land, sometimes reluctantly and against opposition by other claimants.[11] Lately we have seen massive privatization of state property in the former Soviet Union and its former satellites.

What these episodes tell us is that the development of property rights is a social process shaped by the power structures in society. The Viking society was ruled by chiefs but was democratic to a degree. The American settler society was without chiefs. Government in the United States has prided itself of being by the people, of the people, and for the people, and in a society so founded it was undoubtedly difficult to oppose rights developed and recognized by people at large. The power structures shaping the privatization of Soviet state property have been described in many books.[12] That process has not been a pretty sight, which does not necessarily mean that other avenues were open. The privatization process in Russia is probably best understood as a process where the old elite associated with the command and control economy was deliberately replaced with a new elite based on private property, in order to cement the new power structures. Ironically, the same individuals appear to have been equally comfortable within both elites.

What these recorded events do not tell us about, however, is the primeval development of property rights. Settlers of new lands take their history with them, and they can be expected to apply those parts of it which they find appealing and practical. Private property certainly existed in Viking-age Norway, and the European settlers in North America knew of property, even if they themselves had none in the Old World; perhaps for that reason it was all the more important. The reformers of Russia got their property notions from the capitalist world. Individual property rights are not known in all primitive societies, but power and influence are not necessarily equally distributed. It is possible to view economic history as a victory march of property rights. Gradually private property has become accepted as the normal order of things, supported by those who wield power in society. In primitive societies the club or the spear decided; in more developed and ordered societies it was wealth and organization; wealth and custom could command the armies necessary to suppress any rebellion against the prevailing order. But even in present-day democratic societies private property has come to be accepted as the normal order of things, despite being unequally distributed, presumably because it has shown itself to be a more productive way of organization than socialism.

Besides the recorded history of settlements on new lands, we have several examples of how property rights have changed form on land that previously was common. This has happened as a result of technological changes or changes in demand, making it worthwhile to claim individual property rights to previously common land or to change the form of land tenure. In the English enclosures, as a result of a rising demand for corn, common pastures were claimed by individuals, fenced in, and turned into farms. In Scotland, increased demand for wool made it more profitable to turn the Highlands into grassland for flocks of sheep than to extract rents from subsistence farmers and squatters, who were forcibly driven away. In the

1870s, the invention of barbed wire made it possible to contain herds of cattle, making it worthwhile to claim individual property rights to what was previously common grassland in the American West. Let us look a little closer at the English enclosures and the Scottish clearances, because these examples of privatizing the commons are in many ways parallel to what has been happening in the oceans in recent decades.

The English Enclosures

The English enclosures occurred over centuries, beginning in the fifteenth century or possibly earlier.[13] Enclosure involves dividing common fields into private plots and fencing them off. In medieval England there were two types of common fields. One was the open fields under tillage where crops would be rotated on a certain schedule. Each household had its own parcels of the common field, not one but many and scattered. After the crops had been harvested the field was opened as a common pasture for the animals kept by the villagers. This kept the villagers to a common schedule of sowing and harvesting. Advantages of consolidating holdings into a contiguous plot for each owner was one reason why enclosures came to be preferred. Consolidation permitted the tillage of "balks," land which was left unused between the plots. It also permitted economy of trans-portation, described as follows by a seventeenth-century observer: "Dis-order appears thereby, the intermixt and dispersed lands, lying here one and there another. . . . So likewise in carriage of manure and harvest stuffe, and also other carriages, the labour is lost, which might be saved, if each man's land lay together."[14]

Another source of productivity gain lay in being released from the stric-tures of a common schedule, which could hinder experimenting with a new crop rotation. Turnips, for example, were harvested later than other crops and would have interfered with common grazing in the open field. And some farmers did not adhere strictly to the common schedule; they plowed their fields late and trespassed the already sown fields of others to the latter's detriment.

The other type of common fields was the common pasture where the villagers grazed their animals. The incentive to put too many animals on the common pasture has been put succinctly by Hardin.[15] Peasants will continue to add cattle on the common pasture as long as the animals survive and produce enough meat or milk for the owner of an additional animal. But each additional animal eats grass that other animals could have eaten. Beyond a certain point the entire stock of cattle produces less meat or milk than fewer animals could have done. The peasants may end up with a cattle herd that just barely survives, and by their injudicious actions they might push themselves and their families to the edge of existence or even beyond. Some commons were "stinted," i.e., the number of animals each could keep was limited, but even about those commons there are stories to the effect that they were overgrazed.[16]

Enclosure of the common pasture involved dividing it into private plots which subsequently often were tilled and used for crop production. Enclosure could also involve reclaiming unused land such as forests and fens and turning them into arable land. This was the result of marginal land acquiring value because of rising demand for agricultural products, through population growth and industrialization. The earliest enclosures (mid fifteenth and early sixteenth centuries) were, however, a different kind of response. In the fourteenth century, England was ravaged by the Plague, as was the rest of Europe. The population diminished and labor became more scarce. Fields under corn were converted to pastures for raising sheep. This is a bit similar to what happened much later in the Scottish clearances for other reasons than declining population. Concern arose over depopulation in some places:

The towns go down, the land decays;
Of corn fields, plaine lays;
Great men maketh now-a-days
A sheepcot of the church.[17]

From these times we have the adage "enclosures make fat beasts and lean poor people," an indication that all did not share equally in the gains from enclosures.[18] There are reports of uprisings against enclosures in the sixteenth century.[19] One author wrote of "the poor who, being driven out of their habitations, are forced into the great towns, where, being very burdensome, they shut their doors against them, suffering them to die in the streets and highways."[20]

Privatized and enclosed common grazing fields were turned to crop cultivation or better utilized for livestock. Claims of enormous productivity gains can be found in the literature. For an example, see table 1.1. This

Table 1.1

Example of productivity changes due to enclosure. Source: Report on enclosures, Committee of the Board of Agriculture, 1794, quoted from Scrutton 1887, p. 121.

	Weight (pounds)	
	1710	1790
Cattle	370	800
Calves	50	148
Sheep	28	80
Lambs	18	50

puts some empirical blood into Hardin's famous fable, but much more than that was involved. Privatization made improvement of land worthwhile. One interesting case is the draining of fens in Yorkshire and Lincolnshire. This was entrusted to Cornelius Vermuyden, a Dutchman. (Where but in the Netherlands would one find an expert on drainage and dikes?) This was not welcomed by the commoners living by these fens, who

led a kind of predatory life, fishing and shooting wild fowl, a lazy, lawless existence, almost in a state of nature. They kept a few geese, some sheep if well off, and perhaps even a horse. They had freedom to range over a large tract of land, which they had hitherto called their own; and any change which would compel a settled and labourous life appeared to them odious, and they opposed it with the vigour that an open air career had given them.[21]

In our day and age these would be called lifestyle arguments. Vermuyden was promised one-third of the reclaimed land for himself, but he had to contend with various kinds of sabotage from the commoners. Similar incidents occurred when the fens in Cambridgeshire were drained.

The early enclosures proceeded by agreement among those who shared the common field. Undoubtedly some of those who "voluntarily" acceded to an agreement had their hands forced by those who had more at stake and were more powerful; indeed there are stories to this effect.[22] The first recourses to having Parliament approve of enclosures were made in order to overcome the resistance of a minority. Were they just recalcitrant, or did they stand to lose from the process? After 1700 the process of having Parliament approve of enclosures became more and more frequent. That enclosures brought considerable economic benefit can be concluded from the facts that recourse to Parliament was expensive and not all petitions succeeded. In addition, hedges had to be planted, and sometimes ditches had to be dug. It appears that the parliamentary procedure provided some safeguard against unfair treatment of those who were affected by the process; however, those who had least wealth also could least afford to plead their case before Parliament.[23]

The increased involvement of Parliament finally resulted in a general enclosure act being passed in 1801. Enclosure commissioners were appointed to oversee all enclosures. One of their tasks was to make sure that the division of the land was fair. This probably strengthened the hand of those who had least wealth; the commissioners investigated each case and held hearings on the spot. Practice could, however, differ. One thorny question was the treatment of those who had no legal titles to the common but had used it nevertheless, living nearby. Sometimes the commissions regarded their interests as legitimate, sometimes not. Increased recourse to the commons by such landless trespassers may in fact have been one reason for enclosure. Some contemporary commentators had harsh words about them: "The men who usually reside near a common are the depredators of the neighbourhood; smugglers, sheep stealers, horse jockies and jobbers of every denomination here find their abode."[24] Not all contemporary observers would have agreed that the poor were well served by preserving the commons. One put it this way: "Where wastes and commons are most extensive there I have perceived the cottagers are most wretched and worthless and accustomed to rely on a precarious and vagabond susbsistence."[25] A description, *mutatis mutandis*, of some open-access fisheries of our time? Today we would call this a poverty trap.

As time went on, considerations other than agricultural productivity came to have a bearing on the enclosure process. The growth of congested cities, and especially London, created a need for open spaces for recreation for townspeople, it being "much better for them to have such places left open to them, than to be shut out and left to no other resource than the alehouse."[26] In the latter half of the 1800s opposition to enclosures grew for precisely this reason, leading in 1865 to the formation of the Commons Preservation Society. We may detect here a certain parallel with latter day developments with respect to creation of exclusive use rights in fisheries, particularly in the United States. As public use aspects of fish resources have become more prominent (recreational fisheries, fish as wildlife) the development toward exclusive use rights has slowed down, these being instrumental for raising commercial productivity but not for other purposes.

The enclosures were controversial; many were concerned that they made the rich richer and the poor poorer, although few probably doubt their wealth enhancing effect. It has been alleged that the productivity gains resulting from the enclosures enabled England to withstand the vagaries of climate variations much better than the subsistence agricultural economies on the continent of Europe, such as France.[27]

Like the present-day process toward rights-based fishing, the enclosure process had its strange twists and turns and far-fetched arguments. Some felt enclosures would interfere with the fox hunts. One supporting argument was that the hedges would halt the advance of invading armies.[28] Legislators could, then as now, be capricious and throw out proposals for no apparent good reason. The first general enclosure bill was proposed by the Committee of the Board of Agriculture in 1794 and passed by the House of Commons but was thrown out by the Lords, who detected an anti-church attitude in the bill because it proposed a conversion of tithes.[29] Perhaps it was a vicarious argument; for some reason, some "private interests" were against the bill.

The Scottish Clearances

The clearances in Scotland in the late 1700s and the early 1800s were more brutal than the latter-day English enclosures. The clearances did not require any legislation by Parliament; it was enough for the landowners to discontinue their leases and serve eviction notices on their tenants. Some of the estates in Scotland were almost like mini-kingdoms within the United Kingdom; for a time, the estate of the Duchess of Sutherland exceeded a million acres (over 4,000 square kilometers).

Yet the Highland clearances abound in ironies and paradoxes.[30] The Scottish Highlands, even as late as the early nineteenth century, were a Malthusian world where population growth was tempered by recurrent famines. The rural idyll was punctuated by barbarous codes and customs. There is a report of a Highland widow remembering her first two husbands with pride; they had both died an honorable death, being hanged as cattle thieves.[31] The only way to improve the lot of the Highlanders was through raising the productivity of the land. This required radically new ways in using the land and a resettlement of many if not most of its inhabitants, either in new towns on the coast or by emigration. What made the problem particularly acute was that the rising demand for wool in the wake of the Industrial Revolution, together with a new breed of sheep, had made the Highlands much better suited for grazing sheep than for any other activity. For this the smallholders and cottagers of the Highlands would only be in the way.

The grandest experiment in reshaping the Highland economy occurred in Sutherland in the 1800s. The Duchess of Sutherland had married Lord Gower, who inherited one of the greatest fortunes in England and was to become known as Lord Stafford. His wealth was brought to bear on the transformation of Sutherland, but with modest returns.[32] The Duchess planned to relocate her redundant Highlanders to towns on the coast and to have them engage in fishing and manufacturing. The intellectual fathers of this proposition were two entrepreneurs from the south shore of the Moray firth, William Young and Patrick Sellar. Young was a successful industrialist, Sellar a lawyer educated in Edinburgh and influenced by classical Scottish political economy of the eighteenth century. Their ideas would today be called a win-win proposition. The rents of the land could be increased many times over by replacing the Highlanders with sheep. The lot of the Highlanders could, in turn, be improved by engaging them in more productive work, be it fishing or manufacturing.

This was, needless to say, a good thing for the Duchess to believe in. She wanted greater income from her estate, and she desired to improve the lot of her subjects. But when push came to shove, the Highlanders proved recalcitrant, preferring to go on living in their old and established ways. Many would not leave their cottages voluntarily, and in the end some were evicted through burning their dwellings to the ground. Lives were lost; Sellar was in fact charged with manslaughter and the gallows cast a shadow over him until he was acquitted.

Sellar leased a large tract of land from the Duchess of Sutherland and became one of the most successful entrepreneurs in Britain at the time. His thoroughness in clearing the land of its tenants was undoubtedly boosted by his personal need to get them out of the way; he had borrowed a large sum of money at a high rate of interest and could not afford to lose much time in waiting to stock his farm with sheep. He saw himself as a major improver, a man of a new and enlightened age, and had little but contempt for the Highlanders' archaic and unproductive ways:

... they lived in family with their piggs and kyloes, in turf cabins of the most miserable description; spoke Gaelic only, and spent their time, chiefly, in winter converting potatoes and a little oatmeal into this manure; and in summer converting this manure again into potatoes.[33]

His views on what now would be called restructuring were unambiguous:

... if Ground be unsuited for tillage, it is wrested from the possession of 50 ignorant persons, who keep upon it, God's plants in a state of decay, and His creatures in the most abject and pitiable state of misery; and it is put into the possession of one man; who, if he mean to pay his increased rent, must, and he will guide the whole, to health, happiness and prosperity—the former rude occupants draining together into villages, and they and their descendants prosecuting those branches of industry, for which this particular district or country where they happen to be situated, is best adapted.[34]

Sellar's outspokenness and success turned him into a celebrated hate figure for all rural romantics and defenders of the old ways and culture of the Highlands. He has been the bogeyman of innumerable stories and movies, and occasional profaning of his grave continues to occur to this day. His antagonists were a sundry lot. Karl Marx, in *Das Kapital*, used the clearances in Sutherland as an outstanding example of oppression and exploitation by agrarian capitalists. One of Sellar's main antagonists and defenders of the old ways of the Highlands was Stewart of Garth, who for a time was a general in the British army and who had much appreciation for the warrior tradition of the Highlanders. He has been described as a relatively liberal leader of his troops, having "found it necessary to have only two shot and one hanged in his entire regime with his regiment."[35] Stewart died as governor of Santa Lucia, the profits from his family's slave-operated plantations in the West Indies, as the Caribbean Islands were then called, having been insufficient to secure his finances.

In 1852, after the clearances in Sutherland had passed their zenith, the second Duchess of Sutherland (daughter of the first Duchess and Lord Stafford) hosted Harriet Beecher Stowe, the author of *Uncle Tom's Cabin*, on her anti-slavery lecture tour of Britain. The irony was not lost on Karl Marx, who dabbled as a correspondent for American newspapers at that time. It is possible to find quotations from the first Duchess of Sutherland which indicate an attitude toward the Highlanders probably not very different from that of benevolent and enlightened slave owners in the antebellum South. The Highlands used to be a fertile ground for recruitment of soldiers, but when her tenants responded unenthusiastically to the recruitment drives to raise an army against Napoleon the Duchess remarked, disappointedly, that the people "need no longer be considered a credit to Sutherland, or any advantage over sheep or any useful animal."[36] In clearing her land for sheep farms, she did not want to disband the peasantry or drive them to emigration, but "a proper degree of firmness" would be necessary, and the clearances would at any rate provide an opportunity to "get rid of them in case[s] of bad conduct."[37]

Mrs. Stowe responded vigorously to the attacks on her association with the Duchess, referring to "ridiculous stories" about the Duchess and adding that "one has only to be here, moving in society, to see how excessively absurd they are."[38] Mrs. Stowe's movements in society did not extend to the Highlands, but no doubt she had hospitality to pay.

Despite their infamy, the clearances in Sutherland were by no means the worst.[39] Many of those who were cleared came out better, and certainly their descendants gained. Some settled on the coast and became fishermen; others emigrated to better climes and more fertile soils in Canada and Australia, where, in breaking new ground, the aboriginal inhabitants were cleared away.

But sheep farming in the Highlands came and went. After a few decades, the Highlands became more valuable as a hunting ground for the rich and noble, and sheep gave way to deer. And the crofters gained political clout. They formed a political party of their own and gained great leverage in a hung parliament in 1885. This resulted in the Crofter Act of 1886 by which crofter tenants gained a permanent and inheritable possession of their crofts. The Crofter Act did little, however, to alleviate poverty in the Highlands, and it is regarded by some as having obstructed progress by freezing the tenancy structure of 1886.[40] However that may be, it is an illustration of how the evolution of property rights is shaped by the power structures in society. In the early days the crofters of the Highlands had little influence in the British Parliament, and the Scottish lairds could ultimately rely on state power to evict their tenants. In a society where the common man had a greater influence, such as the United States of the same period, informal rights of tenants and squatters would have had a better chance of being recognized. Whether more democratic governments generally are more conducive to a definition and enforcement of property rights promoting economic growth and welfare than less democratic ones is a different question, however.

Enclosures on the Sea

The enclosures and the clearances and the debates they gave rise to have many features in common with enclosures on the sea in our times. Instead of hedgerows and ditches, boundary lines have been drawn on sea charts. But inside the exclusive economic zones of individual countries, fish are often still a common resource for the residents of the countries involved. The tragedy of the commons has certainly not been absent from those commons, not necessarily by way of reducing the total catch of fish but rather through increasing the cost of fishing and foregoing a more productive use of labor and capital. Industry and governments have realized, sometimes belatedly, that there are gains to be made from privatizing the commons. The way this has taken place is through exclusive fishing rights, although usually not of a territorial kind. The debate about this privatization and its effects harks back at what happened during the enclosures and the clearances. Not everyone gains in equal measure. Some perceive a loss and oppose privatization. Recourse has been made to parliamentary processes to seal agreements within the industry (e.g., the American Fisheries Act). Part-time fishermen have sometimes been "cleared" away without any compensation (New Zealand). Academics, journalists and other commentators, not necessarily having any personal stake in the fishing industry, have become engaged in a debate on the perceived unfairness of the private use rights. And there is little doubt that the exclusive use rights have resulted in major gains in efficiency. In fact we only have to substitute the words "fish" and "fisheries" for "land" and "agriculture" to make Gonner's words from 1912 an up to date comment on present-day fisheries policy[41]:

Of even greater importance was the change whereby agriculture from being a means of subsistence to particular families had become a source of wealth to the nation . . . From this point of view the retention of a system which withheld land from its best use was an obstacle to general progress only to be defended by arguments equally applicable to any improvement or invention in a productive process.

The enclosures and clearances and similar episodes have given rise to what has come to be known as the economic theory of property rights.[42] This theory holds that property rights become established when the benefits of claiming and enforcing them surpass the costs thereof. How the benefits and the costs count depends on the power structures in society at each time and place. The losses of the Highlanders were not a factor to be reckoned with in the clearances. But even if all costs and benefits are appropriately accounted for, establishing property rights over what has become a scarce resource due to technological development can increase the wealth of society. Private property rights will ensure that the object is put to the most productive use and that its future productivity is not jeopardized through excessive use in the short term, provided the rights are good for the long term. One need look no further than to the difference between the way people take care of owner-occupied versus rented dwellings to see how property rights incentives work.

The oceans were the last commons to be enclosed. For the most part, the fish resources of the oceans used to be common property, to be enjoyed by anyone who had the audacity and the equipment necessary to go after them. There were good reasons for this. Fishing technology was for the most part so primitive that the effect of fishing was of limited consequence; the abundance of fish was influenced more by environmental fluctuations in the ocean than any human activity. It may now seem strange, but even as recently as the late 1800s biologists could debate whether fishing had any effect at all on the abundance of fish stocks. The leading British biologist at the time, Thomas Huxley, was firmly of the view that fishing did not much matter.[43] The benefits of claiming property rights to fish in the sea were thus of doubtful value. Furthermore, the costs of claiming such rights were high. Navies would have had to be assigned to monitor foreign fishing vessels and to drive unauthorized vessels away. The attempts by King Charles I to take a cut from the Dutch fisheries off the coasts of England and Scotland were not particularly successful.[44]

Even so, exclusive rights to fisheries were claimed in various places around the world. The lords of feudal Japan gave exclusive fishing rights to designated villages.[45] This, however, was for the purpose of appropriating the fruits of other people's work rather than for managing fish stocks as scarce resources. Interestingly enough, these feudal rights have evolved into exclusive rights now held by fishermen's cooperatives in the inshore fisheries of Japan, an arrangement apparently quite successful in preventing overfishing and, in some cases, enhancing the productivity of the fisheries through improving fish habitat. In the Pacific there existed in various places exclusive rights to fishing spots.[46] To what extent this had the effect of preventing overfishing, consciously or not, is uncertain; rather than the fish being scarce it may have been the case that certain spots had better fish aggregations than others so that good location and not fish was a scarce commodity. Such traditional rights have tended to become eroded as cash economies have developed in these places.

Even in Great Britain there were attempts at claiming exclusive rights for the crown to fish out to a distance certainly exceeding 3 nautical miles but otherwise not clearly defined. This is a bit ironic, given the role Britain later played in arguing that everything outside 3 nautical miles was high seas with free access to anyone, but boils down essentially to the elementary fact that nations' positions on international law are simply projections of their national interests. The reason for this claim was perhaps not so much a concern that fish stocks were being depleted by foreign fishermen as a desire to take a cut from the value that they were wresting out of the seas around Britain and to weaken a competitor and a potential enemy. In the 1600s, Charles I, Charles II, and Oliver Cromwell tried to impose fees on the Dutch herring fleet off the coasts of England and Scotland.[47] They fought three wars with the Dutch, partly over the fisheries issue. Their attempts at establishing a British herring industry came to little, however; the Dutch had superior technology, better access to markets, or whatever that made their fisheries profitable but the British not. For hundreds of years the Dutch sent a veritable armada of vessels to the British Isles for fishing, mainly for herring. In those days the herring fisheries were an important industry. It has been alleged that the herring fisheries were the main source of wealth for the Hanseatic League (Baltic Herring) and the Dutch Provinces of Holland and Zealand in the 1400s and the 1500s.[48] But the herring fisheries had their ups and downs; the fishery off Scania in the Baltic collapsed in the 1400s, and the herring off Bohuslän, since 1658 a province of Sweden, disappeared in the late 1500s and was not seen again for 70 years. Were there environmental factors at work, or was the fishing technology at this early age capable of engineering a stock collapse, as happened with the herring stocks in the Northeast Atlantic in the 1960s? It is too late now to tell.

As the fishing technology improved, particularly with the invention of the steam trawler in the 1800s, the signs that fishing could indeed harm the productivity of fish stocks became clearer. By the early 1900s fisheries biologists in Europe generally agreed that fishing did indeed affect fish stocks. The inadvertent fishing moratoria in the Northeast Atlantic brought about by the two world wars provided further evidence that it could do so in a harmful way; fish catches improved and the fish stocks bounced back after the wars were over.[49] The idea that open access to the sea outside 3 miles was harmful for the fisheries gained ground. A book by Thomas Wemyss Fulton, lecturer at the University of Aberdeen, published in 1911 argued forcefully that the 3-mile limit gave insufficient protection to fish stocks. The book cited cases where English or Scottish fisheries regulations had pertained over areas further out than 3 miles and had in fact been applied to domestic fishermen while foreign fishing vessels could fish up to 3 miles, because of the emphasis Britain put on the 3-mile limit internationally. This gave rise to the possibly first incident of fishing under a flag of convenience; in the early 1900s, trawlers from Hull fished in the Moray Firth in Scotland with impunity under the Norwegian flag, avoiding a Scottish ban on trawling in the firth.

Gradually, certain countries began to claim a wider fisheries jurisdiction than 3 miles. Many countries in fact never accepted the 3-mile limit, at least not for fisheries. Norway and Sweden claimed 4 miles, and Norway enclosed all seas inside its archipelago along the coast and closed two wide fjords: the Vestfjord between the mainland and the Lofoten islands and the Varangerfjord in Finnmark. This was contested by Great Britain, and the case went to the International Court of Justice in the Hague, which ruled in Norway's favor. Iceland also wanted a similar limit and to close off the Faxa Bay, an area with good fishing. Needless to say, the impetus behind these claims was the growing realization that fishing was having an adverse effect on the fish stocks and that foreign fleets had to be displaced to make room for more domestic fishing. After World War II, a process got going by which the international law of the sea would be changed beyond recognition.

Thus, what we in effect have here is a case that fits the economic theory of property rights perfectly. As a resource becomes more scarce, the benefits of claiming property rights exceed the costs of doing so and the time is ripe for enclosure. But the enclosure process is shaped by the way people and states arrange their relationships. The enclosure of the seas was different from the English enclosures and the Scottish clearances. If for nothing else, it had to be because this involved interaction between sovereign states and not one between subjects of a single state.

2 The International Law of the Sea

International law is a set of rules which nations have come to agree, explicitly or tacitly, is in their mutual interest to follow. There is no legislative assembly providing international law, and there is no international police force that can enforce it. There is an international court of justice whose verdicts provide precedence for later cases, but countries must agree on submitting their disputes to that court.

In the process of shaping international law, some nations have always been more influential than others. In practice the leading military powers in each time period have shaped international law according to their interests. This is still largely the case, even if deliberations in international forums have come to play a role our ancestors could only dream of. The rules governing the uses of the sea have therefore been shaped by the most important naval powers. Perhaps the most succinct statement of the forces governing the international law of the sea was made by President Franklin D. Roosevelt in an answer to an impromptu question at a press conference on September 15, 1939: America's territorial waters extend "as far as our interests need it to go out."[1]

Over time the interests of the leading naval powers with respect to the law of the sea have shifted. There have been periods when attempts were made at claiming exclusive rights to vast tracts of the oceans. In earlier times this concerned rights of navigation. The purpose was to monopolize trade for the benefit of particular countries. The most famous attempt is probably the division of the oceans between Spain and Portugal in the 1500s, immediately after the great discoveries. The two countries claimed exclusive rights of navigation on the newly discovered seas. Portugal got the eastern Atlantic and the Indian Ocean, Spain the western Atlantic and the Pacific. This was logical enough; Vasco da Gama sailed east and

Columbus headed west. The arrangement had the blessings of the Pope. The emerging northern powers, such as England and the Netherlands, would have none of this, and the Iberian powers were unable to defend their claims.

In the 1600s, successive English kings, and Oliver Cromwell, made extensive claims to the ocean. Over an area stretching from Stadt in Norway, and possibly even all the way from the North Cape, down to Cape Finisterre in Galicia, and westward possibly as far as America, all vessels passing by British naval ships were supposed to acknowledge Britain's supremacy on the sea by lowering their topsail and their flag. These rights were claimed right up to the coasts of the continental countries. The British did on some occasions go to great lengths in enforcing these claims, and the other party sometimes proved more than a little recalcitrant. The Swedes were particularly reluctant to make the salute. One case is reported of a Swedish fleet losing 150 men in a fight with the British over this, and the British were not without casualties either.[2] In retrospect it is difficult to see what purpose these extravagant claims had. The British did not try to prevent the passage of merchant ships, except in time of war, and were probably unable to do so. They did, as stated earlier, try to make the Dutch fishing fleet off the coasts of England and Scotland pay rent for their fishing, but with little success.

These claims brought the British into conflict with, among others, the Dutch and the Danes. The Danes, for their part, claimed exclusive rights to the sea between Norway and Greenland. The British fishery off Iceland and their trade with the Icelanders were a thorn in the eye for the Danes, and they tried their best to prevent both. The Danes also tried to prevent whaling by the Dutch in the northern seas, particularly at Spitzbergen, where the Dutch built a whaling station. The name Spitzbergen commemorates the presence of the Dutch in the Arctic, as does the island of Jan Mayen, named after its Dutch discoverer. The Danes controlled both sides of the Øresund, as well as other access routes to the Baltic, until 1658, and made all ships that sailed to and from the Baltic pay a toll.

In the 1600s the Dutch were the most energetic defenders of the freedom of the seas. The Dutchman Hugo de Groot (Hugo Grotius in Latin) wrote a celebrated treatise on the subject entitled *Mare Liberum*, with a not so subtle subtitle that translates to "the right which belongs to the Dutch to take part in the East Indian trade."[3] The Dutch as a major trading and fishing nation had a clear interest in keeping access to trade, navigation and fishing as open as possible. The 3-mile limit, for many years virtually the norm for the width of national boundaries at sea, was based on a Dutch idea. More than 100 years after *Mare Liberum* was published, another Dutchman, van Bynkershoek, argued that the territorial sea should be based on the shooting range of a land-based cannon. The estimate of 3 miles has been alleged to originate with an Italian named Galiani.[4] The range of cannons soon enough surpassed 3 miles, but the 3-mile norm nevertheless came to prevail. By the late 1800s, most countries had adopted 3 miles as their territorial limits at sea. It was vigorously supported by Great Britain, which by that time had far surpassed other countries as a naval and a trading power and which also had developed important fishing interests in distant waters. The 3-mile limit was the norm until after World War II, but the norm weakened with the waning of Britain as a world power and with the technological progress in fisheries and oil extraction that made it increasingly obsolete.

The Truman Proclamations and the 200-Mile Limit

The decisive impetus for the revolution in the international law of the sea that occurred in the latter half of the twentieth century came from two proclamations made by US president Harry Truman immediately after the end of World War II. In one of these, Truman proclaimed that all resources on and underneath the seabed on the continental shelf of the United States were the property of the US government. In practice this meant oil resources first and foremost, oil having become a vital source of energy and having become possible to extract from offshore fields at moderate sea depths. The second proclamation concerned fish. Here the president was satisfied with reserving the right to proclaim conservation zones but stopped short of proclaiming exclusive rights of extraction for the United States. Fish was not then, any more than now, of strategic importance.

But in other parts of the world these proclamations were seen in a different light. One of these was Iceland, then as now critically dependent on the fish resources around the island. The Icelanders saw no reason why the rules applied to the resources in the water column above the continental shelf should be any different from those applied to the resources on and underneath the shelf itself. In 1948 the Icelandic parliament passed a law claiming a right to control fisheries in the waters of the continental shelf, which in those days was usually taken to mean the extension of the land mass out to a depth of 200 meters. Similar claims were put forward by Argentina, Panama, and Mexico shortly after the Truman Proclamation. On the Pacific side of South America, Chile, Peru, and Ecuador also laid claims to fish resources off their shores. Unlike Iceland and Argentina, these countries have a very narrow continental shelf, and they exploit fish stocks high up in the water column over depths much greater than 200 meters. To incorporate these resources they claimed jurisdiction stretching 200 nautical miles from the coast. Thus was born an idea which later developed into the 200-mile exclusive economic zone that we are familiar with today.

The origin of the 200-mile limit is somewhat bizarre. It can be traced to a desire by the Chilean whale oil industry for protection against foreign whaling. During World War II, Indus, a company in Valpara'so, developed a new method to process whale fat. The company feared that it would not be able to compete against the whaling industry in other countries once the war was over and sought ways to reserve whaling off the Chilean coast for itself. The legal experts consulted looked for precedents in international law but found little other than a declaration from a meeting of the foreign ministers of the American republics held in Panama in 1939. This meeting had declared a zone of neutrality around the coasts of the Americas south of the US-Canada border. (Canada was already a party in the war, together with Britain and most of its dominions.) In this zone the warships of the belligerent nations were prohibited from any hostile activities, in order to protect the trade and other interests of the American republics. The zone was demarcated by lines drawn between specified coordinates, and its width varied between 300 and 500 nautical miles. The 200 miles have been alleged to be derived from an inaccurate map published in a Chilean magazine, Semana Internacional, in January 1940, indicating that the neutrality zone was about 200 miles wide off the coast of Chile. Indus would have been happy with a much narrower zone of 50 miles or so, but for this the lawyers found no precedence.[5]

Figure 2.1 shows the official map of the neutrality zone.[6] The critical coordinates are marked by crosses and the zone by straight lines between. None other than President Roosevelt himself is said to have marked the coordinates and drawn the lines in his White House office.[7] Despite

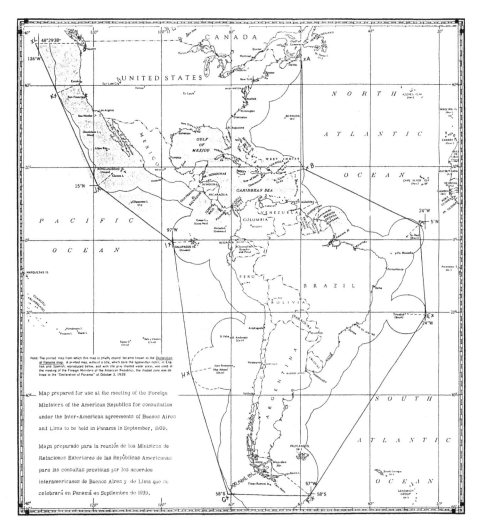

Figure 2.1
Official map of the neutrality zone declared at the 1939 meeting of the foreign ministers of the American republics in Panama. Source: Foreign Relations of the United
States: Diplomatic Papers, Department of State, United States of America, 1939
series, volume 5.

generous help from a librarian at the Biblioteca Nacional in Santiago, the author has been unable to locate the map in Semana Internacional. A stylized map of the neutrality zone, called the Hemispheric Safety Belt, can be found in a classic text on American diplomatic history.[8] The stylized map shows a generally narrower zone than the accurate map, but whether it was this map or something similar which gave rise to the idea of the 200-mile limit is an open question.

An alternative theory on the origin of the 200-mile limit is that it was based on the Humboldt Current. Hollick (1977) dismisses this as an after-the-fact rationalization. If true, this would not be the only example of using ocean currents as a rationale for territorial limits at sea; no less an authority than Thomas Jefferson argued for basing the territorial sea limits of the United States on the Gulf Stream.[9]

The UN Conferences on the Law of the Sea

After World War II, fishing technology improved continuously and the world catch of fish increased steadily by about 6 percent per year up to the early 1970s. Parallel to this the pressure on the fish stocks also increased. The international disputes over fish stocks increased in intensity, but so did also the attempts to defuse them and to reach internationally agreed solutions. The United Nations held three conferences on the law of the sea in the period 1958–1982. The first of these, held in Geneva in 1958, managed to reach agreement on the rules governing exploitation of resources on and underneath the seabed. In essence this conference codified the Truman Proclamation, recognizing the ownership of coastal states of the resources on and underneath their continental shelf. This was an important step, without which the development of offshore oil extraction might have been very different.

The 1958 Geneva conference failed, however, to reach agreement on jurisdiction over the living resources of the sea. At that time the major fault lines were between those who wanted to preserve the old 3-mile limit and those who wanted a wider jurisdiction, at least for fish resources. A compromise with an extension to 12 miles and recognition of historical rights was put forward but failed to reach the required two-thirds majority, both at the 1958 conference and at the second UN Conference on the Law of the Sea 2 years later.

The evolution of technology put the unresolved issues in the law of the sea under an increasing pressure. Technology moved on two fronts: the pressure on fish stocks continued to intensify, and the mining of mineral nodules from the deep seabed came within reach. This stimulated two very different agendas. First, the pressure from coastal states wanting to establish ownership over the fish resources off their shores continued to intensify in tandem with the advances in fishing technology and increasing pressure on fish stocks. Second, opposition arose against the possibility that certain states or private companies would get hold of the mineral resources of the deep seabed. This opposition came both from idealists such as Ambassador Arvid Pardo of Malta, who in 1967 made a celebrated speech at the UN about the deep sea as a common heritage of mankind, and from spokesmen for the developing countries[10] who feared that the rich countries would gain hold of these resources for themselves by virtue of their technological superiority. Furthermore, some mineral-exporting countries saw deep-seabed mining as a potential threat to their exports and therefore wanted a say in regulating this activity.

Pardo's speech is often cited as the impetus for the third UN Conference on the Law of the Sea. The immediate reaction was the establishment of the UN Seabed Committee, which dealt with issues related to mining on the ocean floor outside the continental shelves. Antagonism between rich and poor nations soon developed in this committee. This division was to mar the proceedings of the third UN Conference on the Law of the Sea, ultimately prompting a few of the rich nations to vote against the emerging Law of the Sea Convention. In contrast to the preparatory committees for the previous conferences, the Seabed Committee did not succeed in putting on the table any concrete proposals. After several years of unfruitful deliberations it was decided to convene the third UN Conference on the Law of the Sea. The conference came to consider a much broader range of issues than seabed mining; all outstanding issues related to the uses of the sea were put on its agenda. In addition to deep-seabed mining it considered fisheries, pollution, navigation, research, and issues regarding the delimitation of the continental shelf raised by the ongoing advances in the technology for offshore oil extraction. Paradoxically, this conference was to result in the most comprehensive enclosure ever of common resources, despite Ambassador Pardo's eloquent speech about our common heritage and how the grabbing of the vast common ocean space by individual countries would increase the disparity among nations.

Originally scheduled to end in 1974, the conference spanned a period of 10 years (1973–1982) with intermittent sessions. The major stumbling block was the regime for deep-seabed mining, which is ironic, because this was hardly an economically viable proposition at the time and is still only at a preliminary and experimental stage, more than 20 years after the conference was concluded. The deliberations on this issue got infected by ideology; this was the time when there was much talk of a new economic world order, a time when the price of oil rose to unprecedented levels and many were concerned about scarcity of strategic minerals. The conference was also enormously big and correspondingly unwieldy; practically all the member states of the United Nations participated in it. The conference was in many ways unprecedented and invented its procedures in part as it went along; because of its size it had to be broken up into smaller working groups, and when these became too large, informal groups emerged under the leadership of strong personalities. To avoid majorities consisting of nations without much weight to throw around in world affairs a consensual approach was adopted. The operational meaning of this had to be clarified; what constitutes consensus, when does an amendment lack support, and so on and so forth. It is easy to ridicule many aspects of the procedures of the conference, as some authors have done, but national legislative assemblies are not void of empty gestures and ceremonies.[11]

The countries participating in the third UN Conference on the Law of the Sea had widely varied interests in fisheries. Some fished mainly in waters near their coasts and were interested in reserving these resources for themselves. Others sent their fleets far and wide, some even all over the world (the Soviet Union in particular and to a lesser extent Japan). The initial proposals on sovereign states' fishing rights reflected this diversity of interests. From the beginning there was, however, a strong undercurrent favoring increased rights of sovereign states in the waters off their coasts, at the expense of distant water fishing nations and nations with access to the sea but a short coastline or encircled by states bordering on the open ocean (Sweden versus Norway and Denmark, for example).

There were three principal ways in which the interests of the coastal states could be favored: (i) extending the jurisdiction over the continental shelf to the waters above the shelf, (ii) defining access rights to fish stocks with reference to in which state's waters they originate or mainly are found, and (iii) establishing fisheries jurisdiction over a certain territory irrespective of ecological or topographical factors, such as 200 miles from the shore. From the point of view of efficient control of fish stocks the second alternative would have been most logical. This was, in fact, the gist of the second proposal on fisheries that the United States put forward at the conference. But, true to its national interests, the proposal exempted so-called highly migratory species. This, practically speaking, meant tuna, which US-based vessels were involved in fishing both in US nearshore waters and distant waters.[12]

Despite its apparent sensibility, there would most likely have been problems with putting a proposal like that into practice. The jurisdictional issue would have been a critical point; i.e., would the coastal state have had the power to enforce rights over its stocks at any distance from its shores? If not, the coastal states' rights would have been largely worthless. In many cases it would have been difficult to ascribe a fish stock to a single coastal state's waters so that two or more coastal countries would have had to cooperate in managing the stock, but that is not different from what obtains under the arrangement that came to prevail. In any case the proposal got little support, but the idea of ownership belonging to the coastal state, or the state where a fish stock originates, lives on in the articles of the Law of the Sea Convention pertaining to anadromous species (species like salmon which are spawned in rivers but migrate to sea for feeding, returning as mature fish to spawn). Fishing for these species on the high seas is forbidden, unlike other migrating species.

The idea of giving coastal states jurisdiction over the waters above the continental shelf would have had an element of functionality similar to assigning ownership rights to fish stocks to the coastal state(s) where they are mainly to be found, as the migration of bottom dwelling fish is restricted to the waters above the continental shelf. This would have been logical if the rights to resources on and underneath the seabed had continued to be limited to the continental shelf. That was, however, not to be; continued technological development made it possible to exploit resources under the ocean bottom at greater depth than 200 meters, and so the conference struggled for a long time with defining coastal states' rights under what came to be called the continental slope and rise, the falling off of the continental shelf to the deep ocean floor. The simple idea of a 200-mile exclusive jurisdiction over resources on, underneath and above the sea bottom, quickly gained ground. The second meeting of the conference in Caracas in 1974 showed that this solution had wide support.

The consensus approach of the third UN Conference on the Law of the Sea (in the first two, proposals were adopted if they got two-thirds of the votes) undoubtedly prolonged the conference and accounts for the inconsistencies and lack of clarity marring the text of the convention that was finally adopted; papering over differences of opinion by inconsistencies and ambiguities is a well-known trick to reach agreement. One commentator refers to explaining the obscure in terms yet more obscure.[13] Sometimes this can make laws or treaties utterly unworkable, and it is in any case a gold mine for lawyers, but in other cases it may be innocuous, with the ambiguous paragraphs being largely irrelevant. Despite the consensual approach, a few countries, most notably the United States, voted against the Law of the Sea Convention when the conference was concluded in 1982. The main reason was the rules governing deep-seabed mining, an issue which has largely gone away, partly because deep-seabed mining has turned out not to be profitable or only marginally so, and partly because the rules governing this activity have now been watered down so as better to suit the interests of the United States and other industrially advanced countries.

The Law of the Sea Convention formally became international law in 1994, having by then been ratified by the required minimum of 60 states. (As of late 2003 it had been ratified by 145 states.) Nevertheless, some of the rules it codified became international law *de facto* much earlier, some even before the third UN Law of the Sea Conference had been concluded. This is true, for example, of the 200-mile limit. Already in the latter part of the 1970s many countries established a 200-mile exclusive economic zone. It was, in fact, the United States Congress which jumped the gun by passing the Fishery Conservation and Management Act in 1976, establishing a 200-mile exclusive fisheries zone for the United States. (A 1983 proclamation by President Ronald Reagan turned this into an exclusive economic zone.) Many countries soon followed suit. This action went against the will of the US negotiators at the conference and two important departments of the US government, the Department of State and the Department of Defense, but was nevertheless signed into law by President Gerald Ford, who faced an election campaign and could ill afford to lose any state. (Eventually he lost the election to Jimmy Carter.) The US government saw the Law of the Sea Convention as a package deal; its overriding imperative was freedom of navigation through straits and archipelagos, meaning *inter alia* that submarines would not have to surface. For this the United States was willing to concede jurisdiction over resources. These goals were shared by the other superpower at the time, the Soviet Union. Extended jurisdiction over fish partly gave this game away.

Despite some ambiguities, the Law of the Sea Convention must be hailed as a major achievement. Hopefully it bodes well for future lawmaking in the international arena. Most of human history is a history of ethnic cleansing and rule by the club, the sword, the cannon, the machine gun, and the bomb. Gradually, and with setbacks, we seem to be entering the phase of the rule of law. The UN conferences on the law of the sea are perhaps the most impressive achievements so far of the rule of law in international affairs. The time-consuming deliberations of the third conference had many of the trappings of a legislative assembly, even if the delegates were not elected. Factions were formed, horses were traded, logs were rolled, and delegates could claim to have returned home with barrels of pork. Most important, contentious issues were settled peacefully.[14] That said, there is little doubt that the role of the superpowers at the time was decisive; had they been against the Law of the Sea Convention it would never have evolved into international law. Even if the United States did not for a long time sign the convention it could live with it and in fact implemented some of the rules it laid down. Because deep-sea mining was not an economically viable activity, the critical issues were never brought to the fore, and by now they have been defused by the above-mentioned new rules on deep-sea mining.

The Enclosure of Fish Resources (Not Quite Finished)

As far as the fisheries were concerned, the third UN Conference on the Law of the Sea was indeed a sea change. The 200-mile limit once claimed by a handful of states in Latin America became recognized as an exclusive economic zone in which the coastal state has the right to manage the exploitation of natural resources. For fish resources this right is conditional insofar as the coastal state is required to share with others any surplus that it cannot utilize, but in practice this is a dead letter; it is the coastal state that decides what the total allowable catch of fish is, and there are many examples of coastal states leasing fishing rights to others at their own discretion. For all intents and purposes the fish resources within the 200-mile exclusive economic zone have become the property of the coastal state, and the sovereign rights of the coastal states to manage these resources as they best see fit appear to be universally recognized.

How could this happen? It is one thing to say that the countries with rich fish stocks off their shores had an interest in appropriating these resources but quite another to explain how they got away with it. In earlier times the law would have been made by military power, but here we are dealing with peaceful negotiations where there must be mutually advantageous give and take if a settlement is to be reached. It appears that the countries which were after the fish had a perhaps unique window of opportunity. The two superpowers at the time, the United States and the Soviet Union, had a common interest in preventing "creeping jurisdiction" which might interfere with their interests in unimpeded navigation of military vessels through straits and archipelagos that would become territorial waters if the limits of such waters were extended. To avoid this they were prepared to concede jurisdiction over natural resources within a 200-mile zone or whatever could be agreed upon while restricting the territorial sea to 12 miles, with special provisions for passage through straits and archipelagos. But the exclusive economic zone also owes much to the fisheries interests of the United States. The US was one of the countries that stood to gain handsomely from extending national jurisdiction over fish resources, even if it also stood to lose some interests in distant water fishing, mainly over tuna. In fact the United States tried as best it could to defend its interests in tuna by having it classified differently from other fish as "highly migratory species" and subject to rules of management different from other fish.

The 200-mile zone was also promoted by the fact that the developing countries endorsed this idea, even if many of these countries are either land-locked or have a short coastline. The fact that the land-locked and the so-called geographically disadvantaged countries did accept this new order is surprising, and they did not do so without a fight. The empty phrase that they could share in any surplus of fish which the coastal states could not utilize on their own was designed to placate them. Did they believe it? It is probably one of the examples where papering over differences of opinion with ambiguities turns out to be irrelevant. The definition of surplus is up to the coastal states, which seem to have a large leeway, to say the least, to do so in a way that leaves nothing to other claimants. Was this phrase also harmless? Perhaps not; there are some indications that coastal states expanded the capacity of their fishing fleets so as to be sure that there was no surplus to be shared, and that they went too far in so doing. In fact the world is still wrestling with a legacy of overcapacity of fishing fleets. More important perhaps, the land-locked and geographically disadvantaged states probably had little leverage against the coastal states once a critical mass of the latter were prepared to push for the exclusive economic zone. Needless to say, it is the coastal states which control access to the sea, and it is by trading with the coastal states in one form or another that land-locked states can share in any benefits from the resources of the sea.

The outcome of the third UN Conference on the Law of the Sea stands in a marked contrast to the UN Conference on Straddling Stocks and Highly Migratory Fish Stocks. This conference was convened in 1993 in order to deal with the growing problems associated with fishing on the high seas. The conference did not go for what would seem to be the logical response, namely a further extension of the 200-mile limit. The main reason for this undoubtedly is that the conference was about a single issue on which countries had diametrically opposite positions. On one side were the coastal states which wanted to increase their influence over fisheries immediately outside their 200-mile zone, while on the other were distant water fishing nations who wanted as few restrictions on their high-seas fishing as possible. A compromise had to be somewhere in between. The third UN Conference on the Law of the Sea, by contrast, was about many issues, and the compromise was that different groups of countries got their way on the issues they were most preoccupied with. The coastal states got the 200-mile limit, the superpowers freedom of navigation, and the developing countries the seabed mining regime. No such multi-issue compromise was possible at the UN Conference on Straddling Stocks and Highly Migratory Fish Stocks.

The UN Conference on Straddling Stocks and Highly Migratory Fish Stocks ended by concluding an agreement in 1995, which vested the rights to manage fish stocks on the high seas in regional fisheries organizations.

There are at least two problems with this. First, these organizations do not have the power to exclude anyone from fishing on the high seas. Secondly, these organizations have no jurisdictional power to enforce whatever rules they may agree on. Jurisdiction over a fishing vessel on the high seas is still in the hands of the state whose flag the vessel is flying. If the countries which at a point in time exploit a high-seas fish stock did agree on managing that stock and did so successfully, there is a risk that this would serve as an invitation to some state which is not a member of the club to start fishing the stock, so eroding the gains that the members of the club have achieved.

It is doubtful, therefore, whether this arrangement will result in an efficient exploitation of fish stocks on the high seas. At the time of writing there are stories both of success, albeit limited, and failures. The states fishing in the so-called Donut Hole in the North Pacific, a high-seas area enclosed by the 200-mile zones of the United States and Russia, concluded an agreement on fishing in that area in the 1990s, but not until fishing in the area had come to an end because of declining catches.[15] Agreements on migrating fish stocks in the Northeast Atlantic have been short-lived (e.g., Atlanto-Scandian herring) or not forthcoming (e.g., blue whiting). Agreements on the fishing of tuna have either failed or are largely ineffectual. In cases where fishing vessels from third countries have ignored the regulations put in place by regional fisheries organizations the member countries have tried to apply trade sanctions against them, apparently with some success.[16]

3 Property Rights in Fisheries

To better understand the role and nature of property rights in fisheries, consider what happens when a fish stock is accessible to anyone. This is often referred to as the common property case, although that is not exactly true. A fish stock that is accessible on the high seas is no one's property. A fish stock that is available only for a restricted number of players, such as a stock within a state's exclusive economic zone, is common property for those whom that state authorizes to exploit the stock. The difference need not be of much consequence, however. In great many cases states do not impose very strict access controls on fish stocks within their economic zones, and some have practically no controls at all. The critical point is whether those who have rights of access to fish stocks coordinate their actions so as to maximize their joint returns from the stock, or at the very least get a better overall result than if they did not cooperate. It probably does not take a very large number of players for cooperation to become impossible, but how many it will take will vary with the traditions of the country in question, how well the players understand the situation, their mutual trust, whether they differ greatly in size and other characteristics, and doubtless much else. For all practical purposes there will be little or no difference between free access to fish stocks on the high seas and lax access controls to fish stocks within a country's exclusive economic zone.

Open access to fisheries leads to what has come to be known as the tragedy of the commons, briefly discussed in chapter 1. Peasants holding grassland in common are likely to impoverish themselves through over-grazing. In the worst-case scenario, the productivity of the grassland might be irreversibly destroyed. This is most likely to happen in marginal areas where vegetation fights a precarious battle against a hostile climate. Over-grazing still appears to occur in certain places, such as the Sahel, on the edge of the Sahara, and Lapland, where reindeer graze on slow-growing lichen. Poverty and lack of alternatives exacerbate the tragedy of the commons; a cow that is critical for feeding one's family now will not be given up for more grass in the future. Hence there may emerge a vicious circle of overgrazing and poverty; erosion of grassland may decrease the overall yield from the animals while one more cow may still feed one more family, and so on to a common ruin.

Critics of the tragedy hypothesis have argued that the peasants surely must understand the situation and in their common interest limit the number of animals to what the grassland can support. If the peasants are not too many and their interactions are characterized by mutual trust or dominated by one or a few sage peers, this could happen. Open-access fisheries are, however, typically characterized by a large number of players, and the effect of fishing on the productivity of fish stocks is less visible than the effect of overgrazing on grassland. There is every reason to expect overfishing to occur in open-access fisheries, and numerous studies have in fact shown this to be the case.

Some Simple Fisheries Economics

What, then, is overfishing? Three degrees of overfishing can be distinguished. The most serious one occurs when fish stocks are exploited so intensively that they become extinct. Extinction of fish stocks has rarely happened, however. The second type is the depletion of fish stocks to levels that support a smaller sustainable catch of fish than it would be possible to attain from a more plentiful stock. This can be called *biological overfishing*. The third type of overfishing covers the case when labor and capital and other factors of production used in the fishing industry yield less in economic terms than they could. This can be called *economic overfishing*, and it can occur even if there is no overfishing in a biological sense.

How and why, then, does overfishing occur? Suppose we start exploiting a fish stock that has never been exploited before, and suppose there are no controls on the fishery; anyone can invest in new fishing boats and fish as much as he wants to. In the beginning, while the stock is still plentiful, the boats participating in this fishery will get good catches and do better than cover their costs. Other boatowners, or presumptive boatowners, will observe this and conclude that this is a good business to be in, and more boats will enter the fishery.

The large initial catches and profits will not be sustained, however. A fish stock that has never been exploited can be thought of as being in a natural equilibrium where there is no surplus growth; the growth of the stock, through individual growth of fish that survive and recruitment of young fish, will be just sufficient to compensate for natural deaths. The oceans would not fill up with fish even if we would stop fishing, and unexploited fish stocks have survived over the millennia, so the idea of a stock in a natural equilibrium where growth is balanced by decay is not a bad one, as a first approximation. But there are large year to year fluctuations in stock growth, and there seem to be fluctuations in stock abundance on a longer time scale. In this context we can ignore these complications and think in deterministic terms reflecting average conditions.

To fix ideas, we can use this simple relationship:

Change in stock = Surplus stock growth – Catch.

Surplus growth is the difference between stock growth and natural decay. In the initial situation (pristine equilibrium) there is no surplus growth, so catching fish will mean that the stock is diminished correspondingly. But fish stocks have been exploited since time immemorial without being driven to extinction, so it must be true that some surplus growth emerges as stocks are driven below their natural equilibrium level. Surplus growth could emerge for a variety of reasons. First, fewer fish could mean less competition for food, so that individual fish grow bigger and faster. This effect is most likely for fish in confined environments such as lakes. Second, a higher rate of exploitation means that the age structure of the stock is displaced toward young, fast-growing fish. Third, large stocks of spawning fish could have an adverse effect on recruitment, through competition for food among eggs and fry, but too small spawning stocks are likely to have an adverse effect on recruitment.

If the catch were exactly equal to the surplus growth, there would be no change in the stock; the catch would be sustainable. But this cannot happen in a natural equilibrium, as there is no surplus growth in that situation. In order to accommodate the catches of fish, the stock will have to be driven down below the natural equilibrium level. There is no way we can have both an undisturbed natural equilibrium and a sustained fishery.

A possible relationship between the stock level and the surplus growth of fish is shown in figure 3.1. A stock in a pristine equilibrium produces

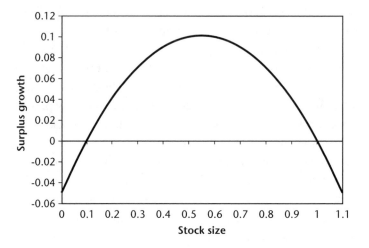

Figure 3.1
A possible relationship between surplus growth and equilibrium stock size. Note two equilibrium stock levels, 0.1 and 1.0. The latter is the natural equilibrium without fishing. The former is a critical threshold level of viability; if pushed lower, the stock would go extinct.

no growth and cannot sustain any catch. An absence of fish, or a stock that has fallen to its lowest viable level, also produces no surplus growth and no sustainable catch. Intermediate stocks produce some surplus growth and are able to sustain a catch equal to that growth, and one particular level produces maximum surplus growth and hence maximum sustainable yield. Note that a sustainable catch could be obtained over a wide range of stock size and that there is a range of sustainable catch levels, from nothing at all to the maximum surplus growth that the stock will support. The notion of sustainability does not get us far toward answering the question of how many fish to catch or what stock level to aim for, except that the catch should not consistently exceed the maximum sustainable, in which case the stock would become extinct.

But let us return to the story of the fishery that has just developed. More boats are being added to the fishery. What happens to the catch? If the catch per boat were to stay the same, the total catch would increase in proportion to the number of boats. More boats would keep on coming in if the catch per boat were to stay the same, because there would be an excess gain over and above costs to be cashed in by new boatowners, unless the market became glutted with fish and the price fell. But a steady increase in the number of boats and a corresponding increase in the catch cannot be sustainable. There is a limit to how large the surplus production of a fish stock can be. The stock would be fished to extinction.

Is this a realistic scenario? Not many fish stocks seem to have gone extinct. The main reason why this has not happened even in fisheries with little or no control over the total catch or fishing activity is that the catch per boat falls as the stock is driven down. The fish become fewer and more difficult to find. Fewer fish get entangled in the nets or bite the hooks, and fewer fish are swept up by trawls. Revenues will no longer exceed costs, and profits may turn to losses. When that point has been reached, there will be no incentive any more for new boats to come into the fishery, and some boats that have entered might leave. Ultimately the number of boats and the stock would settle down to a new equilibrium where the catch is equal to the surplus growth.

The story does not have to end, however, on this happy note. If the stock has the habit of aggregating, it might simply become concentrated in a smaller area as it is driven down. In that case the fish might not be so difficult to find even if there are just a few of them, and as many as before would be caught in nets, in trawls, or on hooks, as long as there were any of them left. Profits would remain positive to the end, attracting more and more boats until the stock had been wiped out. There is some indication that some fish stocks behave this way. One example is herring. The herring stocks in the Northeast Atlantic and the North Sea were almost wiped out in the late 1960s, and what may have saved them was a total ban on catching herring that lasted several years. Herring have the habit of gathering in large shoals, and the catch per boat does not seem to be much affected by the size of the stock. A stock might also be driven to extinction because the catch per boat has not fallen sufficiently to eliminate the difference between revenues and costs until the stock has been driven down below its lowest viable level (figure 3.1).

But what about the equilibrium with a sustainable catch? Is that a happy situation? The equilibrium will become established when the value of the catch per boat has fallen to a level equal to the cost per boat, eliminating the incentive for new boatowners to enter the fishery.[1] The more boats there are the smaller is the equilibrium stock, and since the catch per boat falls as the stock becomes smaller, the catch per boat will fall in the long

term as the number of boats increases. But the long-term (sustainable) contribution of an additional boat to the total catch will fall even faster and
could in fact be negative, reducing the sustainable catches of fish instead
of augmenting them.[2] This would happen if the equilibrium stock lies to
the left of the maximum sustainable yield level in figure 3.1, where the
stock has been driven down to a low and unproductive level. This is exactly
analogous to the overgrazing example discussed above.

The fact that the sustainable contribution of an additional boat to the
total catch could be negative is a stark illustration of the fact that all is not
well with free access to fish stocks. Each fishing boat represents a commitment of productive resources—manpower, capital, and various imple-
ments—to one particular activity which has the purpose of enhancing our
material standard of living.[3] All these resources can normally be used for
producing material benefits in a different way: capital can be invested in
manufacturing, people can work in factories, and money can be spent on
raw materials for those factories rather than on bait, nets, and other imple-
ments for fishing. A society that uses its productive resources optimally
would allocate them in such a way that it gets the same value from the
last unit allocated to all activities. The last unit of labor would produce the
same value in fishing as in manufacturing and in farming, and capital
would give the same return in fishing as in other industries. But this will
not happen in the open-access fishery. Under those conditions the value
of the catch per boat will be equal to the cost per boat, which necessarily
means that the value of the sustainable contribution of an additional boat
will be less and possibly negative. Given that the cost per boat represents
the value that the resources used by the boat (capital, manpower, fuel, etc.)
could produce otherwise, the boat would not be paying its way in the long
term; it would have been better to use these resources in a different way.

To see more clearly what this means, figure 3.2 is useful. Remember that,
in equilibrium, more boats mean a smaller stock of fish. The relationship
between sustainable yield and the number of boats must therefore be somewhat like a mirror image of the curve in figure 3.1; the catch will increase
with the number of boats up to a point and then fall (the surplus growth
increases as the stock is reduced, and then declines as the stock is reduced
beyond a certain point). If the price of fish is independent of the catch
volume, the sustainable catch value curve will have the same shape as the
sustainable yield curve, so we can without loss of generality set the price

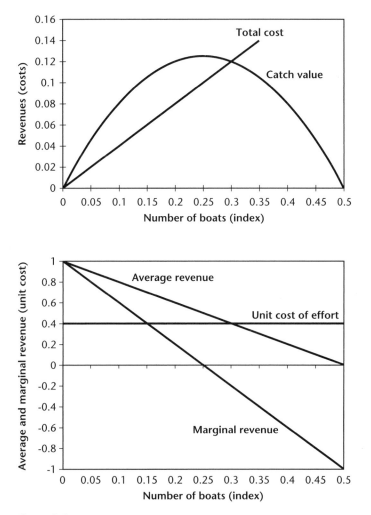

Figure 3.2

Upper panel: How total sustainable revenues and costs depend on the number of boats. The minimum viable fish population is set at zero, and the sustainable catch per boat is assumed proportional to the size of the fish stock. Lower panel: How the sustainable revenue per boat (average revenue) and the sustainable revenue of an additional boat (marginal revenue) depend on the number of boats. The open-access equilibrium is where the index number of boats is equal to 0.3; the maximum difference between sustainable revenues and costs (rent) is achieved with only half as many boats (0.15). Note how the sustainable contribution of an additional boat can be negative even if the sustainable catch per boat is positive and even greater than the cost per boat.

equal to 1. (We can always account in units that have a value of 1.) The lower panel of the figure shows the catch per boat and the value of the sustainable catch contributed by an additional boat. The equilibrium in the open-access fishery will be where the value of the catch per boat is equal to the cost per boat, but to maximize the contribution of the fishery to the material well being of society we would want to be where the additional value produced by an additional boat is equal to the cost per boat.[4]

This reasoning assumes that the cost per boat represents the value that otherwise could be produced by the resources used on the boat. This is by no means unlikely. Boatowners will not invest in fishing unless they get a return equal to what they would get elsewhere, and they will have to pay their hired labor a wage comparable to what they would get elsewhere. There is reason to believe that the return on capital in other sectors represents the value produced by an additional unit of capital in those sectors, and that the wage rate represents the value produced by an additional unit of labor, but it is true that various so-called market imperfections (of which open access to fisheries is one) could cause deviations in other sectors similar to or in directions opposite to what happens in the fishery, so a full assessment of the inefficiency of the fishery can be a complicated piece of analysis.

From figure 3.2 we see that the optimal number of boats is less than the equilibrium number under open access. The productive resources that are tied up unnecessarily in fishing under open access represent a loss of material welfare; these resources could have produced a greater value if they had been used in a different way. In addition, the overfishing may be so severe that the fishery in fact produces a lesser value than the maximum sustainable with a greater use of resources, as shown in figure 3.2. This would amount to biological overfishing. But economic overfishing need not imply biological overfishing; as the reader can verify, a high enough cost of boats would produce an open-access solution on the left hand side of the maximum sustainable yield, but there would nevertheless be a loss involved because the last boat is contributing less than what it costs.

This mismatch between the economically optimal and the open-access solution occurs because the effect of the stock level on the productivity of the fishery is not taken into account. This can be seen as a consequence of the fact that no one owns the fish stock. An owner of the fish stock depicted in figure 3.2 would have an incentive to maximize the rent from the stock, either by fishing the stock himself or by renting out the right to do so.[5] The upper limit to the rental payment he could extract from the stock would be equal to the difference between the revenues and costs of fishing, and a maximization of this would in fact imply that the socially efficient solution would be attained.[6] It is appropriate to emphasize that the reason for this is not that maximization of rent is an obvious social goal. Why would it be? Suppose the owner of the stock is some individual or company awash in cash from before; what social purpose would it serve to increase the wealth of that entity even further? But it just so happens that maximization of the rent is a means to achieve maximization of the material welfare obtained from the resources at society's disposal. If it is judged unfair that the rents from the fishery accrue to the owner of the stock, there are ways in which a part or maybe practically all of that rent can be taxed away for the benefit of society at large.

Property Rights

The establishment of the 200-mile limit went a long way toward establishing states' property rights to fish stocks. Examples of such stocks are the Icelandic cod stock and several groundfish stocks off the coasts of the United States, all of which are fully contained within the 200-mile exclusive economic zone of each country. More commonly, however, fish stocks migrate across national boundaries. In such cases, two or more countries have common ownership over the stocks. This is the case, for example, in the North Sea. In the Barents Sea, Norway and Russia share the stocks of cod, capelin, and other fish.

Then there are cases where fish stocks migrate into the high seas, the area outside 200 miles. Distinction is often made between straddling stocks and highly migratory fish stocks. Straddling stocks are such as stay mainly within the economic zone of one or more countries but migrate occasionally, or to a limited extent, out of the zone. The cod stock in the Barents Sea is one example; it is partly accessible in an area outside the Norwegian and Russian economic zones called "The Loophole." Another is Alaska pollock, some of which are found in the "Donut Hole" between the American and Russian zones in the Bering Sea. A third is the cod stock on the Grand Banks of Newfoundland. Parts of the Grand Banks protrude out of the Canadian 200-mile zone.

Highly migratory stocks, on the other hand, migrate over long distances and can be found successively in the economic zones of several countries as well as on the high seas. The most important of these stocks are tuna of various kinds. Finally there are stocks exclusive to the high seas. Such stocks are found on ocean ridges outside 200 miles. Orange roughy probably is the most important of these.

Despite the fact that all stocks are not entirely enclosed within the 200-mile limit, there can be no doubt that this new order of the sea fundamentally changed the institutional framework for fisheries management. The exclusive economic zone provided a legal framework that made it possible for one state, or for two or more states sharing a stock, to limit access to the stock. Without this framework, rights-based fisheries management systems, some of which will be discussed in later chapters, would not have been possible. Rights must be embedded in a legal framework specifying what they are, and they must be embedded in a judicial framework establishing when violations have occurred and what punishments should be applied to those who break the rules. This last function must be embedded within some sovereign state's judicial system; there exists no international police force to enforce international law, and it is not likely to come into existence any time soon.

The main weakness of the 200-mile zone is that it does not establish property rights over fish stocks as such, except when stocks happen to be enclosed within the zone. Not surprisingly, the establishment of the 200-mile limit encouraged fishing on the high seas, and the amount of fish caught in that area could be as much as 10 percent of the world catch.[7] Fishing of some straddling stocks outside 200 miles has caused considerable furor from time to time. Fishing of turbot by Spanish vessels on the Grand Banks in the mid 1990s led to hot pursuit by Canadian Coast Guard vessels and a media circus in New York's East River. Allegedly illegal nets used by these vessels to catch "baby fish" were put on display in front of the UN building, where the conference on straddling stocks and highly migratory fish stocks was taking place. Fishing by Icelandic vessels in the Loophole in the Barents Sea troubled relations between Norway and Iceland for several years in the 1990s. Open access to straddling stocks undermines the management of such stocks in two ways. First, it represents an uncontrolled taking and weakens the incentive to conserve stocks to a greater extent than the quantity fished would lead one to believe.

Second, it undermines the legitimacy of restricting access to the coastal country's own fishermen, as they can see foreigners enjoying free access outside 200 miles.

The coastal state has *de facto* property rights to a stock enclosed by the 200-mile limit. It can regulate access to the stock as it sees fit, for whatever purpose it deems appropriate. There are, however, many ways in which this can be done. One would be to establish property rights to fish stocks vested in private individuals or companies or in publicly owned corporations. There are certain advantages in doing so rather than having the government control fisheries; politics has a habit of creeping into such processes and interfering with maximization of value. With fish stocks privately owned (or owned by a profit-maximizing public corporation), their management would become profit oriented and market driven, which would provide incentives to utilize them optimally, as discussed above. But, as also noted above, certain market imperfections external to the fishery could prevent profit maximization by stock owners from reaching a socially optimal solution. One such imperfection would occur if private firms were less willing than warranted, from a social point of view, to sacrifice current and unsustainable profits for sustainable future profits. This would lead them to deplete fish stocks further than warranted, in order to obtain short-term unsustainable profits. There are several possible reasons for such undue impatience. A private firm engaged in fishing faces a greater risk than society at large and might have difficulties in diversifying its risk exposure by operating in financial markets. A firm in such a position would have to borrow money at a high rate of interest including a risk premium and would thus need a correspondingly greater cash flow to satisfy its creditors or investors. A firm depending on borrowed money could also face bankruptcy in case it proved unable to pay interest and amortization on its loans as scheduled; the creditors could be less than impressed by sustainable benefits promised for the future. But even if the solution optimal for the private owner would fall short of a socially optimal solution, it would certainly be preferable to open access.

There are few if any examples, however, of fish stocks having been made private property or even handed over to public corporations. One reason may be ideological reluctance to give up public ownership of what is regarded as a common property of the nation, even in countries that otherwise base their economic system on private property rights. Furthermore, many fish stocks migrate across national boundaries at sea. Two or more states would therefore have to cooperate in establishing and enforcing private property rights to such stocks. This is likely to be a tall order. Yet another reason is that some fish stocks migrate out of the 200-mile zone and into the high seas. Property rights to such stocks would be tenuous and of negligible value for the most migratory stocks.

In cases where fish stocks are confined to a single jurisdiction, there are other obstacles to establishing private property rights over them. Some are of a practical nature; others are associated with the transfer of wealth that might accompany such rights. Some fisheries are quite profitable, potentially at least, and so establishing ownership rights over the stocks they exploit would confer substantial economic benefit upon those who got the ownership rights. Even competitive auctioning of rights to fish stocks would probably not be a very successful method for capturing the future rents associated with exploiting the stocks, due to the long time horizon and the uncertainty involved. Private property rights to fish stocks might also be difficult to manage and enforce. Fish stocks typically migrate over large areas so that the rights would have to be geographically extensive. Furthermore, the habitats of fish stocks overlap, and different stocks are sometimes fished indiscriminately by the same fishing gear. Ownership rights defined in geographical terms might therefore not be effective, unless covering very large areas including a complex of stocks. The biological interaction among stocks is also an argument for extensive geographical rights covering the habitat of all interacting stocks. If ownership rights to stocks were to provide appropriate incentives to protect interrelated stocks and optimally utilize their productivity, they would have to cover the entire complex of interrelated stocks.

For these reasons, it is unlikely that private property rights to fish stocks will be widely applied to deal with the problems arising from common property. On the other hand, exclusive use rights, despite being more circumscribed, could go a long way toward solving the problems arising from common property. Exclusive use rights do not confer any rights to the fish stocks as such, only rights of access or utilization. Several countries have, in the wake of the establishment of the 200-mile exclusive economic zone, established fisheries management regimes based on such rights.

A fisheries management system based on exclusive use rights instead of property rights to the stocks themselves implies a division of tasks between

governments and the fishing industry. In this scheme of things governments are the guardians of the viability and productivity of fish stocks
through setting an overall limit to how much can be fished from each particular stock in any given period, either through an overall catch quota or
through an appropriate limit on fishing effort (the sum total of the activities of the fishing fleet). In the economist's ideal world, governments
would determine the overall catch limit, or the effort limit, on the basis
of rent maximization, as discussed above with the aid of figure 3.2. Today
the governments of countries sharing a fish stock negotiate how much can
be taken from the stock or how much effort can be applied in the fishery,
and governments with jurisdiction over entire stocks within their 200-mile
zone set an overall catch quota or an upper limit to fishing effort. The
economic optimization applied in this process is however quite crude or
nonexistent. The limits on the total catch or fishing effort are set on the
basis of recommendations by fisheries biologists. These recommendations
are based on biological and not on any economic reasoning, and governments usually do not follow them to the letter, for economic and social
reasons. Whether the deviations from the biologists' recommendations
move us any nearer to an economic optimization is another issue and is
probably not the case. Governments do not necessarily have incentives to
maximize economic efficiency in any particular industry.

With governments setting the limits on fish catch or fishing effort, it is
up to the industry to determine how the fish is taken, with how many
boats, etc. The important thing to note is that use rights can be defined in
such a way that they provide two crucial sets of incentives: (i) incentives
for the industry to conduct its operations efficiently and (ii) incentives for
the industry to use whatever influence it may have on government decisions in such a way that the productivity of the fish stocks is adequately
preserved, similar to what property rights to the stocks themselves would
accomplish.

There appear to be three major ways in which exclusive use rights to fish
resources can be defined: (i) rights to catch a certain quantity of fish, (ii)
rights to own and to operate fishing vessels, and (iii) territorial use rights.
There is a certain degree of overlapping between territorial use rights and
the first two. Both rights to land a certain quantity of fish and rights to
use fishing vessels can be geographically circumscribed. In fact this is
always done in the sense that a right to fish only applies within a certain

state's exclusive economic zone, but the right could be further restricted, and this is in fact often the case. Territorial use rights could, however, be defined without any further restrictions, in which case it would be up to the right holder to decide how much he takes of the fish in his area and in what way. The comments below pertain to such "pure" territorial rights.

Rights to Catch a Certain Quantity of Fish

The form of exclusive use rights that appears most promising for achieving an efficient exploitation of fish stocks is *individual transferable quotas*, usually referred to as ITQs. This form of exclusive use right is probably the one that has been most widely applied since the 200-mile economic zone came into being.

An ITQ is a right to catch a specific quantity of fish from a given stock within a given time period. The individual quotas can be determined either as fixed quantities or as shares of the total catch permitted in any given period. This system is therefore appropriate where a fish stock is protected by setting an upper limit to how much can be caught from the stock in any given period. This kind of conservation policy has been extensively applied since the 200-mile zone came into being. Due to the uncertainty in estimating the size of fish stocks, a total quota could be inappropriate and could endanger the existence of the stock.[8] Other methods are not, however, faultless. Controlling fishing effort amounts to an indirect control of the total catch and could be preferable to controls based on imprecise stock estimates if the relationship between catch per unit of effort and actual stock size were stable and independent of environmental conditions, but that is not necessarily the case.[9]

In a fishery controlled by an overall quota, be it the best way to proceed or not, one will go a long way toward maximum economic efficiency by dividing the overall quota into individual allocations and make them transferable in the long and the short term. Such quotas provide strong incentives to maximize rents in the fishery and to utilize fish stocks efficiently, for any given time profile of total catch determined by the government's resource management policy (and which could be inefficient). Quota holders have incentives to maximize the value of the catch they are allowed to take and to minimize the operating costs for taking it. If the quotas are valid for the long term, they also provide incentives for quota holders to invest optimally in fishing vessels.[10] If the government's conservation policy is based on clear rules that do not change over time in an unpredictable way, the quota holders will be able to make rational predictions of how large catches they can expect to be allowed to take in the future. Transferability of quotas is, however, of critical importance to achieve efficiency. If quotas can be bought, sold, and leased freely they will most likely end up with those who can pay the highest price for them, which normally are those who can utilize them most effectively.

ITQs are particularly promising as a vehicle to rationalize a fishery which has developed overcapacity, as they can be implemented in a way that amounts to an industry-financed buyout program. Allocating quotas among those who at a certain point in time are engaged in the fishery and allowing them to trade quotas would lead to the more efficient quota holders buying out the less efficient ones. Hence the overcapacity of the fishing fleet would gradually be eliminated. In some fisheries where ITQs have been implemented the consolidation of quota holdings has been rapid. Cases in point are the Australian southern bluefin tuna fishery[11] and the surf clam fishery in the United States, the latter to be discussed in greater detail in chapter 7.

ITQs are not, however, without problems. The most serious problems, from the point of view of efficiency, are associated with monitoring and enforcement. The incentive for the individual quota holder to cheat and not to report his catch is obvious. ITQs are therefore not likely to be a good option in small scale fisheries where fish landings could easily become a part of the informal economy (say, when fish are landed on the beach and sold directly to consumers). ITQs also strengthen the incentives to throw away less valuable fish at sea, as the quota holder obviously wants to maximize the value of landings obtained on the basis of his quota. Even if much ink has been poured to deplore this activity it is not clear how serious it really is. Fishermen are not likely to waste much effort to catch fish in order to throw it away; such activity does not provide much income. A fish which it is not worthwhile to bring ashore would in many cases be a product which the ultimate buyer is not willing to pay much for and had perhaps better be thrown overboard. It is possible, however, that some fish which it would be worthwhile to bring to market would be thrown away to make room for still more valuable fish within a given quota. This is wasteful and, if unreported as it usually is, creates problems for the fisheries biologists whose task it is to assess fish stocks and come up with recommendations about the total allowable catch and who need accurate and timely catch statistics. Discarding of fish as a result of a quota regime is particularly likely to arise in fisheries where different species of fish are being caught indiscriminately. The fisherman may have a quota for some species but not for others, and the easiest solution may be to throw away any fish which one is not authorized to take, even if it could fetch a high value in the marketplace.

In the cases where ITQs have been put in place they have usually been allocated free of charge to boatowners in the fishery concerned, mainly on the basis of their recent catch history. One reason why quotas have been given away is that in many if not most cases they have been introduced at a time of crisis, with fishing capacity exceeding the capacity needed to take the permitted catch and many and perhaps even most firms in the industry operating at a loss. The ITQs have then been an instrument for industry-financed rationalization where the most efficient firms have gradually bought out less efficient ones. Over time rents have emerged in the industry and become capitalized into a market value of ITQs. The ongoing controversy over the ITQs in Iceland, to be discussed in chapter 6, is in part over this. Some quota owners have made a windfall gain from having got fish quotas for free and then sold out of the industry. Although the value of quotas is a result of rationalizing the industry and not something being taken at anybody's expense, this unfolding of events is seen by many as unfair.

To deal with the problem of windfall gains, the quota rent could be confiscated in one way or another. One way is to levy a special tax on holding quota, which would reduce the market value of quota allocations. Some countries have in fact elected to do so. In Namibia the industry pays fees for fishing licenses and quotas amounting to about 8 percent of the landed value, a percentage that was higher in the past.[12] In the halibut fishery in the Canadian Province of British Columbia there is a resource rental fee of 310 Canadian dollars per tonne of quota, and, in addition, the industry pays the costs of managing the quota.[13] Iceland has recently introduced modest resource rentals which will not do much better, however, than cover the management costs of the fisheries. But some countries have elected not to apply resource rentals. New Zealand abandoned its original plans for resource rentals as a part of a dispute settlement with the industry. In the United States it is illegal to claim resource rentals, but the industry can be charged for management costs of ITQ systems.

Another possibility to impose resource rentals is to auction quotas. This is in fact likely to be the most successful method of rent capture. The amount paid for the quotas would be determined by the willingness of the firms in the industry to pay for them, and only the firms themselves know how much they are willing to pay. Auctioning quotas may, on the other hand, dilute the incentives for investing appropriately in fishing fleets and preserving fish stocks. Long-term tenure of quotas is desirable for providing incentives to invest appropriately. With a quota allocation that is valid for at least the lifetime of a fishing vessel the quota holder either knows how much he is allowed to fish or can make rational predictions about this (when quotas are defined as shares of a variable total catch) and determine his fishing capacity accordingly. Long-term tenure of quotas is essential for generating an interest among quota holders in preserving the productivity of the fish stock, as the value of a long-term quota allocation depends on how well the fish stock is being managed. Shortening the period for which quotas are valid would reduce their value and dilute its dependence on the expected future productivity of the stock. Lowering the market value of quotas by resource rental fees is likely to reduce the interest of the industry in promoting stock management, as it would have less at stake.

One possibility of reconciling rent capture through an auction system with long-term validity of quotas would be to auction annually a certain proportion of the quotas. The quotas could in principle be of infinite validity but would depreciate in a geometric fashion at a certain rate. The fishing rights system established in Estonia in early 2001 was an example of this; 10 percent of all fishing rights (defined as quotas, fishing days or fishing gear units, depending on the type of fishery) were auctioned off annually, with the remaining 90 percent allocated on the basis of recent catch history.[14] In 2003 this system was discontinued, however, under pressure from the industry.

There are reasons both for and against taxing away resource rents in fisheries and other natural resource industries. Economists classify rent taxes as "good" in the sense that they do not lead to a change in the supply of labor or investment, unlike taxes on income from labor and capital. There is thus an efficiency argument for using such taxes rather than other taxes. The challenge lies in correctly identifying the tax base (the rent). Tax codes on petroleum extraction, for example, usually do not fully succeed in this. There is reason to believe that taxes on rents in fisheries would be easier to design in this respect. Such taxes could be levied on the quota holdings themselves, or a portion of the quota could be put up for auction each year. The quotas would nevertheless be valuable for the prospective buyer as long as the quota fee did not exceed the rent produced by the quota.

An argument against rent taxes is that they would simply be a painless source of income for governments, which would use the money wastefully rather than reduce other taxes and weigh carefully, on the basis of need for public services versus private goods, how large the public sector should be. Some argue that such taxes would retard economic growth, as more of the rent would be invested profitably if it remained in the hands of the quota owners. Then there is a stewardship argument against taxing away the rent. The rent remaining in the hands of the quota owners will be capitalized into a market value of the quota. This value will depend on how well the fish resources are managed, or are expected to be managed. In order to protect this value, the quota owners will have a collective interest in promoting good management by lobbying for a cautious setting of the total catch quota (which does not prevent each individual from having an individual interest in cheating the system through misreporting and highgrading, much as the individual members of the Organization of Petroleum Exporting Countries have an interest in exceeding their production quotas even if they have a common interest in sticking to them in order to keep the oil price up). The Icelandic boatowners' association (LIU) seems to have adopted that attitude after the individual fish quotas in Iceland became valuable. In earlier years they usually protested against catch quotas being set too low, against the advice of fisheries biologists. Since the quota system was made permanent, they have accepted the advice from fisheries biologists and sometimes argued against politicians who have been willing to set quotas higher than the biologists recommend. Taxing fishery rents would reduce the value of the quotas and so weaken this common interest in good resource management.

Rights to Own and Operate Fishing Vessels

Another form exclusive use rights may take is licenses to own and to operate a fishing vessel with certain specifications for a specific purpose.

Since fishing vessels come in many different shapes and sizes, it is obviously necessary to exercise some control over the technical specification of the vessels for which licenses are issued; if not this will not be a very effective control of fishing capacity. (One case of a lax control is the surf clam fishery in the United States.) Furthermore, since any given fishing vessel can normally be used for fishing different stocks and there is a need to protect each fish stock individually, it is also necessary to specify what type of fish the vessel is allowed to catch. Finally, since fish stocks may vary considerably over time due to environmental fluctuations, it will in most cases be necessary to control the use of any licensed fleet from time to time. Even if the fishing capacity of a given fleet is appropriate in normal times, the fish stocks may from time to time be in a bad shape for reasons that have nothing to do with their exploitation. In such times all the capacity of an otherwise optimal fleet will not be needed, and using it to its fullest extent could endanger the fish stock. The best examples of fish stocks that vary enormously for environmental reasons are so-called small pelagics (fish that live close to the surface of the sea), like sardines, anchovies, herring, and capelin. When an El Niño event occurs, there is very little anchovy off the Peruvian coast, and it would be risky to allow the entire Peruvian fleet to fish at full capacity during such events.[15] Likewise the stock of Barents Sea capelin falls to a very low level under certain oceanographic conditions, and it has been found necessary in such cases not to allow any catches of capelin.

Control of fleet capacity and fishing effort provides only an indirect control of the catch from any given stock of fish. If the fishing capacity of the fleet depends on uncertain factors such as weather conditions, a variable distribution of fish in the sea, etc., this approach to controlling the stock could be very imprecise, with a high risk that the conservation goals would not be attained. There are, however, circumstances where effort control could be better than catch quotas to control the stock level. The size of fish stocks is not easy to assess, and estimates thereof typically have wide confidence limits. If the catch per unit of fishing effort depends on the size of the stock in a reliable way it could be a better estimator of the stock than other methods. Controlling the stock through catch quotas based on stock assessment could then be inferior to controls based on the observed catch per unit of effort. In a number of cases the catch per unit of effort has been an unreliable indicator in this regard, however.[16]

There are other problems with vessel licenses and effort control. The capacity of a fishing vessel is a multi-dimensional variable which it is difficult to control in detail. Experience shows that fishermen expand the capacity of their vessels along uncontrolled dimensions, adding fishing gear if there is no gear control, increasing engine power if they are free to do so, etc. Their ingenuity knows few limits. Dutch fishermen are reported to have circumvented a limit on the power of the main engine by restraining it and installing auxiliary engines which could supersede the "main" engine's power.[17] The phrase "capital stuffing" has been coined to describe this phenomenon; a vessel hull has certain similarities with a Thanksgiving turkey, which can be stuffed with various goodies to enhance its qualities.[18]

Furthermore, technological progress may over a relatively short period increase the capacity of the fishing fleet way beyond what was previously deemed adequate. Over the years there has been considerable technological progress in fishing by way of electronic fish finding and positioning equipment, computer controlled fishing gear,[19] better quality ropes and nets, mechanical hauling instead of hauling by hand,[20] etc. Naval architects have also been very clever at changing the design of fishing vessels and packing more power into technical specifications such as maximum vessel length, etc.[21] Finally we may note a paradox. The more successfully the fishing capacity is controlled the more the industry may be locked in a straitjacket preventing technological progress and the potential gains that go with it. Technological progress often requires a change in the design of fishing vessels, but such changes might run afoul of specifications deemed necessary to keep the capacity of the fleet under control.

One advantage fleet and effort control may have over quota control is easier monitoring and enforcement. Fishing vessels can be counted and measured, and their movements can be tracked, if necessary by a satellite monitoring system. When fish quotas are difficult or costly to monitor, a fleet and effort control could be a better option despite its disadvantages.

Fleet and effort controls are used in many countries. In the Faeroe Islands there is a system of tradable fishing days, but no quotas on fish catches. In Norway a concession is required for fishing vessels above a certain size. Interestingly this system is combined with a quota control system where vessels with concessions are allocated fishing quotas each year. The quotas cannot be leased, but the quota allocations are tradable through buying and decommissioning vessels with quota allocations. The allocations acquired in this way can be stacked on another vessel and kept for a certain number of years. Not surprisingly the value of quota allocations are reflected in prices of fishing vessels over and above what they would fetch as means of production only. One case, widely reported in the Norwegian media in the fall of 2002, involved a boat with a fishing concession being sold for 80 million kroner, stripped of its concession, and sold back to the previous owner the next day for 10 million kroner.[22] Calculating the value of the fishing right is fairly easy in such a case. When fishing concessions are given for free to boatowners, as the case was in Norway and many other places, they get a windfall gain in much the same way as when ITQs are allocated for free. This windfall gain is cashed in by the "first generation" who got the rights for free; for those who come later and have to buy their way in the value of the fishing right is an entry cost on par with other costs, except that they may hope to get it refunded when they in turn sell out of the industry.

Territorial Use Rights in Fisheries (TURFs)[23]

A third type of exclusive use rights involves rights to fish within a given territory. Such rights could be limited to the use of certain types of gear or to certain species of fish, as the conservation goals would presumably be stock specific. The efficiency of this type of use rights depends critically on the migrations of the fish. If a fish stock stays entirely within the assigned territory, the TURF will essentially amount to an ownership of the stock. The implications of stock ownership for optimal utilization were discussed above. For stocks that migrate in and out of the assigned territory the property rights over the stocks will be diluted or in effect non-existent.

An interesting question arises if the entire habitat of a stock is divided between TURFs held by different individuals, firms or organizations. Would they manage to achieve a cooperative solution maximizing the joint benefits from the stock or would they end up in a situation where both of them are worse off, like the two delinquents in the famous Prisoner's Dilemma? The answer depends on the number of agents involved and the differences among them in terms of fishing costs and alternative opportunities. Some attempts have been made at applying game theory to problems of this kind.[24] The results show that a cooperative solution is not impossible but by no means guaranteed. Christy reports a case where a number of Japanese fishermen's cooperatives dealt successfully with the management of migratory fish.[25] The 200-mile exclusive economic zone is in fact a TURF; a country has exclusive use rights to fish within its zone and shares the ownership of a fish stock that migrates into another country's zone with that country.

In Japan the inshore fisheries are managed by TURFs. These rights are usually assigned to fishermen's associations and not to individuals or firms. There exists considerable literature on these rights and associations in the English language, and there are indications that this system is successful.[26] TURFs are also being applied in the Philippines. TURFs are often seen as a vehicle for implementing community-based fisheries management, on which more will be said in the next chapter. A community would then be given management authority over a sea area in its vicinity, as is the case with the Japanese fishing rights. It has been argued that this may be the only way to deal with the problems posed by small scale fisheries; not only would it be impossible to monitor individual quotas in such fisheries, but enforcing a license limitation regime for a large number of small boats might also be difficult. Under such circumstances, TURFs might be the only option left. Such devolution of management issues to the local level might also save administrative resources, usually a major constraint in poor countries, and also appear more legitimate in countries with an open or latent conflict between the center of government and the rural areas where fishing takes place. But in order to be at all effective, the TURF arrangement would have to control access to the resources; somehow the "community" with access rights to the fish stocks will have to be circumscribed. In traditional TURF systems in Asia and the Pacific the access right has typically been prescribed through kinship, apprenticeship, caste, religion or ethnicity,[27] but these approaches do not coexist easily with modern concepts of non-discrimination and equality of opportunity. These traditional systems have typically come under heavy strain from population growth, new and more effective technologies, and cash economies opening up a new outlet for what used to be subsistence fisheries.

Beyond Private Property Rights

I began this chapter by discussing the inefficiencies arising from common property. As the reader will have noticed, the arguments against common property and for individual rights are arguments based on exploitation. The purpose of commercial fishing (recreational fishing is a different case) is the provision of material benefits, mostly for human nutrition (directly and indirectly in the form of feed for animals and farmed fish), but also materials for industrial purposes (the latter used to be the main purpose of whale hunting). As far as commercial fishing is concerned, fish stocks have a value purely as a source of such benefits. Owners of fish stocks will preserve them for the future because they can be a lasting source of such benefits, and also because the fish will probably be cheaper to catch if they are taken from a plentiful stock.

From this it follows that preservation of fish stocks for the future is somewhat coincidental. One can easily construct theoretical cases where it would make most sense, from an economic point of view, to fish a stock to extinction as quickly as possible, because its growth rate is too low compared with the return on alternative assets. In a case like that, an owner of a fish stock would find it attractive to "take the money and run," i.e., deplete the fish stock and invest the money in whatever would give him a higher rate of return. This conclusion does not, of course, sit easily with those who value fish stocks as living resources irrespective of their value as sources of nutrition or materials. The role of private property rights in that context is highly limited; private ownership of stocks would not necessarily give them any protection. There is no conflict with use rights such as ITQs, however, provided the total amount to be caught is adequately set to ensure that fish stocks are protected and maintained at the desired level.

Economically optimal extinction of fish stocks is likely to be an exceptional case, due to generally high growth rates of fish stocks. Neither governments nor fishing industry circles anywhere, as far as I know, have suggested that stocks be fished out because of too low growth rates. Nevertheless, fishing and environmentalism do not always coexist easily. Environmentalists are likely to want larger stocks than the fishing industry and to be more concerned about the risk of stock collapse through overfishing. Environmentalists might also want to catch less fish of a given stock in order to provide forage fish for other stocks not necessarily of any commercial interest, and to restrict fishing that incidentally catches fish and other wildlife. We have seen examples of this in the campaign against fishing with drift nets, and in restrictions on the Alaska pollock fishery for the sake of sea lions.

More fundamentally, the management institutions adequate for environmentalist purposes are very different from those adequate for the provision of material benefits. For the latter, private ownership of fish stocks is likely to be desirable, as has already been argued. Environmentalists, however, are interested in fish stocks not for the material benefits they provide but as a part of the living environment. This approach implies an existence value of fish stocks, a value that benefits everyone, if not in an equal measure; you and I can both simultaneously enjoy the knowledge that a fish stock exists, and my joy over this is not at your expense, unless you are in the fishing industry and would want to catch more fish. Such goods which each and everyone can enjoy simultaneously without interfering with one another are called collective goods, and to ensure a sufficient provision of such goods we need collective institutions such as governments and their agencies, not private property rights which are useful mainly for exploitative values.

The attitude regarding fish as primarily or even exclusively as wonders of nature is a part of an attitudinal change to nature, involving a greater store being set on pristine land otherwise transformed through strip mining or oil extraction, forests otherwise chopped down for timber or paper, and wild animals earlier seen as threat to farming or even to humans going from place to place.[28] This attitudinal change is particularly advanced in the United States and other rich, industrialized nations. It is tempting, to say the least, to see this change as a result of the enormous and still ongoing increase in productivity that has taken place since the beginnings of the Industrial Revolution. This increase in productivity has affected the primary industries such as forestry, farming and fishing no less than manufacturing industries and mightily reduced the number of people employed in these occupations. These increases in productivity have made it possible for manufacturing industries and, later, services of many different kinds, to grow and contribute to an enhanced well being. A side effect of this change is that the outlooks and attitudes of people engaged in industries which no longer account for a large part of the total employment count for less and less as a part of public opinion, for better or for worse. Attitudes and outlooks are probably to a large extent shaped by how people live and what they do for a living. Love of animals dangerous to humans seems directly proportional to the risk of encountering them on television and not hand to tooth and claw in the wild. In occupations far removed from primary industries or industrial production it may be easy to lose sight of the fact that human civilization ultimately derives its amenities from nature and exists only because we humans have learned the laws of nature and turned them to our own advantage; because we have changed nature beyond recognition in the cities where we live and on the farms where we produce our food. Provision of food remains an essential basis of human civilization regardless of how many people are needed to provide it. The fisheries of the world are decidedly less important than agriculture for human nutrition but we would probably not be well advised to try to do without food from the sea altogether, and for some people that is simply not an option, particularly in the poor countries of the world. There are also indications that fish is a healthy supplement to our diet.

Much of what environmentalists contribute to the public debate about fisheries is clearly rooted in an attitude toward marine life as something valuable in itself and not as a source of material benefits in any sense. Sometimes the environmentalist attitude toward fish and fisheries, and toward nature in general, betrays a charming naiveté. The aquarium in Newport, Oregon—for a time home to Keiko, the most famous and expensive killer whale in the world—presents sharks as the scavengers and garbage collectors of the ocean, mostly attacking sick and dead animals, and "humans only when they mistake them for food." At other times we encounter ignorance if not deliberate misinformation. In the aquarium in Monterey, California, we read that the Peruvian anchoveta was fished to a virtual extinction in the early 1970s. As of June 2002 there was still no notice of the fact that the anchoveta stock recovered in the 1980s and bounced back with full force in the 1990s (figure 3.3). The Monterey aquarium also issues a "seafood watch" card advising against eating certain types of fish, *inter alia* "Atlantic/Icelandic cod," a curious taxonomy to say the least. Perhaps the aquarium does this out of compassion toward the Icelanders, whose economy depends to no small extent on the cod fishery; those who refuse to buy their fish would be making a valuable

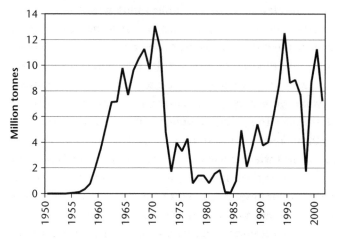

Figure 3.3

Catches of anchoveta, 1950–2001. Source: fishery statistical database, UN Food and Agricultural Organization.

contribution to rebuilding this stock, which would improve the future prospects of the Icelandic economy. In the meantime, the aquarium and the diners they persuade would perhaps be kind enough to put in a good word with the banks that would tide the Icelanders over while they wait for the cod to grow. The aquarium also issues a similar warning for farmed Atlantic salmon. Anyone familiar with this industry would have a hard time understanding how it could possibly be a threat to a "healthy ocean," a cause the "seafood watch" card purports to promote.

4 Toward Individual Use Rights

Exclusive use rights in fisheries were not unknown before the establishment of the 200-mile limit. However, without extended coastal state jurisdiction such rights could only be established in nearshore areas. The territorial rights in the inshore fisheries of Japan have been mentioned as a case in point. These are of ancient origin but were codified in their present form after World War II and assigned to fishermen's cooperatives.

In the late 1960s, individual use rights, in the form of boat licenses, were established in the salmon fishery in British Columbia, the Pacific province of Canada. Only Canada and the United States participated in the Pacific salmon fishery, and countries that could have fished for salmon on the high seas had been persuaded not to do so. A treaty had been concluded between Canada and the United States about how much salmon each could catch offshore, so neither had to race for the fish. The individual boats were, however, in a fierce competition with one another, considerable overcapacity had been built up in the fleet, and as a consequence the fishing season had become much shorter than necessary.

The Canadian government wished to do something about this and decided to limit the number of boats through a licensing system. Licenses were issued to the nearly 6,000 boats in the fishery. There were two classes of licenses. One type could not be renewed and would lose its validity after a certain number of years. These licenses were given to those who did not have a major interest in the fishery. The other type of license, renewable and transferable, was given to those who had a major interest in the fishery, and the idea was that the holders of these licenses, or those to whom they were transferred, would remain in the fishery.

The experience with the license limitation turned somewhat sour, in two ways. First, some people had overoptimistic expectations about the future profitability of the fishery and bought licenses at inflated prices. In the 1970s the salmon fisheries were apparently hit by changing conditions in the North Pacific favoring salmon runs in Alaska but harming runs further south. The second problem came to be known as capital stuffing, a concept explained in the previous chapter. Initially there were no controls on the size of boats and the equipment they could use. The salmon fishermen were still racing for the fish, even if there were fewer boats in the race. Each one had an incentive to increase the power of his boat to win the race. Engines became more powerful, and fish finding and electronic equipment was installed. The cost level in the fishery did not fall much, and the salmon season did not become longer. The problem was that fishermen had no incentives to minimize cost and maximize the value of the fish they were taking. The license limitation scheme in British Columbia became a classic example of how limiting the number of fishing boats through licenses fails to control fleet capacity, fishing effort, and the cost of fishing. The process during the first 8 years after the license limitation was introduced is illustrated by table 4.1, which shows the development of the number of active boats and the nominal value of the fleet from 1967, just before the license limitation program was introduced in 1968, until 1977. While the number of vessels fell by about one-fifth, the value of the

Table 4.1
Number of vessels and their value, in millions of Canadian dollars, in the active salmon fleet in British Columbia. Source: G. Alexander Fraser, Canadian Fisheries and Marine Service, Pacific Region, here quoted from pp. 126–127 of Eckert 1979.

	Number	Value
1967	6,640	71.9
1968	6,603	73.4
1969	6,137	80.1
1970	6,186	88.8
1971	5,806	89.8
1972	5,524	93.7
1973	5,233	93.7
1974	5,207	143.0
1975	5,019	219.1
1976	5,159	241.7
1977	5,292	273.0

fleet increased almost fourfold in nominal terms. Allowing for the decrease in the value of money, about one-third of this could be due to inflation.[1]

Exclusive use rights also emerged in some herring fisheries in the Northeast Atlantic at about the same time. Around 1970 the herring stocks in this area crashed, first the Atlanto-Scandian herring and then the North Sea herring. Fishing for both of these was stopped for a while, but the fleets that used to fish the herring stocks turned to other stocks such as capelin, which previously was virtually unexploited. In Norway a license limitation scheme was put in place for the fleet partaking in these fisheries, as well as individual vessel quotas determined on the basis of the licensed cargo capacity of each boat. Individual vessel quotas were also introduced in Iceland, both for the capelin fishery and for a local herring fishery which had survived the onslaught on the herring stocks in the 1960s. In Iceland these vessel quotas evolved into individual transferable quotas, as discussed in chapter 6, and the same is to an extent true in Norway as well.

In the 1960s, international fisheries organizations, such as the Northeast Atlantic Fisheries Commission and the International Commission for the Northwest Atlantic Fisheries (which has now been replaced by the North Atlantic Fisheries Organization) came up with quota limits for certain stocks. These quotas were for the most part so generous that they were not even binding, being the only limits the member countries could agree on. In any event these organizations were for a time sidelined by the 200-mile limit, which enclosed many of the fish stocks under their mandate. But in any case the idea to control fish stocks by an overall quota had been put forward, giving rise to the idea that this could perhaps be applied at the individual level to attain better economic results.

The idea of individual transferable quotas appears to have been aired for the first time by Francis Christy in a working paper from the University of Rhode Island.[2] His paper clearly and succinctly puts forth the major arguments for individual transferable quotas, as well as anticipating some discussions that were to develop later. He saw "fisherman quotas" as accommodating technological development instead of stifling it, as they would make detailed regulation of fishing vessel design, use of gear, and fishing effort unnecessary. Indeed he put a major emphasis on the freedom of fishermen to choose the most appropriate method of fishing. He correctly foresaw that the initial allocation of quotas would be contentious. He discussed a levy on quotas that might be used to defray the costs of fisheries management and perhaps yield a little extra income for the public purse. He argued strongly in favor of freedom to lease quotas short term, but proposed, without providing any further reason, that they should not be saleable permanently except to the fisheries management agency. But he thought the system would be inexpensive to implement and did not mention quota busting or throwing away less valuable fish at sea.

A few years later, in 1978, there was a seminar on fisheries economics at Powell River, Canada.[3] A few papers addressed the issue of quotas. David Moloney and Peter Pearse, the latter a University of British Columbia professor of economics renowned for his various works on fisheries management, had a contribution about "stinting," an arrangement by which government leases land for grazing, and proposed a similar arrangement for the fisheries.[4] From the paper it is clear that he had fish quotas in mind rather than certain areas where one would have exclusive rights to fish. At a conference held in Australia 2 years later, individual transferable quotas were also discussed.[5] The idea was now afoot and it was gaining support, among economists as well as others.

Individual Rights: Where Does the Initiative Come From?

The story of how the 200-mile exclusive economic zone came about was a story of creation of rights of sovereign states to fish and other resources within a geographic area.[6] The story of the creation of individual rights has parallels but also differences. As was the case for the 200-mile limit, the driving force is scarcity and the desire to make an economic gain by reserving fish resources for some defined group and excluding others.

One difference between establishing individual use rights and the 200-mile limit is that the latter was the result of an agreement among sovereign states. There is no world government which can force any recalcitrant state to abide by what has been agreed in the international arena, in contrast to a sovereign state which can apply its enforcement apparatus against any recalcitrant citizen, or citizens of other states. An international agreement like the Law of the Sea Convention will survive only if sovereign states find honoring it to be in their interest. States, however, come in many shapes and sizes and differ greatly in power. It would therefore be more correct to say that an international agreement like the Law of the

Sea Convention will be honored as long as a critical mass of the most powerful states of the world see it as being in their interest.

Individual rights, on the other hand, must be created and upheld by the
sovereign state on whose territory they are to be exercised and respected.
But there are certain parallels here with the international arena. No new
rights will be created unless a critical mass of the individuals in the society
where these rights are embedded supports that agenda, either because they
gain personally or find it worthy of support for the good of society. What
that critical mass is will depend on the way society is governed. In democratic societies a majority would have to be persuaded, under other forms
of government those who wield sufficient power and influence.

But where is the initiative to create individual rights to come from? To
draw further on the analogy with the 200-mile limit, it could be expected
to come from individuals within the fishing industry. The depletion of
fish stocks translates into falling profits, and maybe losses and ultimately
bankruptcies of firms in the fishing industry. It would certainly be in the
interest of the industry to reverse that kind of trend.

The initiative to create individual rights could, however, come from the
state itself. If the purpose of extending territorial limits at sea is to make
better use of the resources within the enclosed territory, some mechanism
must be put in place to accomplish this. As has been argued above, property rights assigned to individuals or firms would be a way to do so. Establishing individual property rights to fish stocks, or use rights, would be a
logical extension of coastal states' strife to enclose fish stocks within their
territorial limits, as a way of ensuring maximum economic benefit.

In fact we have seen both of these happening. In a number of cases, most
notably, perhaps, in New Zealand and in Iceland, both of which to be
further discussed in later chapters, the respective governments took the
initiative to establish a system of individual transferable quotas. In the
United States, the initiative to establish individual use rights has come
through a joint effort of groups within the industry, a government agency
(the National Marine Fisheries Service), and the fisheries management
councils. While it is true that the federal government has seldom been
directly and actively involved, as the governments of Iceland and New
Zealand have been in their respective cases, the National Marine Fisheries
Service and the management councils represent the government and the

public interest. What is surprising, however, is that governments of coastal states the world over have been less enthusiastic than one might expect to establish individual use rights to fish resources, given the strong thrust for the 200-mile zone. In the concluding chapter we discuss some possible reasons for this.

In the light of the economic theory of property rights, we would expect that players in the fishing industry would themselves take the initiative and persuade governments to establish individual property rights, for the benefit of those who happen to be in the industry at the time. This has indeed occurred, but less frequently, perhaps, than one would expect, given the substantial gains involved. But all use rights systems that have been established were, as far as this author is aware, put in place with the support of a critical mass of the industry, and attempts by governments to do so against the will of a critical mass of the industry have failed.

There certainly are cases, however, where the industry has taken the initiative to establish exclusive use rights such as ITQs. Sometimes this has happened in a "greenfield" situation; i.e., where a fishery has recently been developed and the pioneers have realized that all rents in the fishery would be eroded, and probably sooner rather than later, unless the fishery were closed to new entrants. One such is the wreckfish fishery off the southeastern coast of the United States. But in most cases the initiative has arisen in response to a growing pressure on the resources and an ever tighter regulation of the fishery to safeguard the resources, both of which have led to falling profits in the industry. In order to avoid the erosion of profits the industry has approached the government and asked for secure use rights. Cases in point are the Alaska pollock fishery and the trawl fishery for Pacific whiting.

Other interesting examples of industry initiatives come from Canada. Canada, despite its active role in promoting the 200-mile limit, does not have a clear national policy on establishing exclusive use rights, unlike Iceland, New Zealand, and the Commonwealth of Australia. Individual quotas, sometimes transferable and sometimes not, have emerged in three fisheries on the Pacific coast of Canada over the last 10–15 years. The initiatives have come from the industry in conjunction with civil servants who have concluded that such regimes would be preferable to the traditional controls of fishing time, design of boats, the gear used, etc. In the Atlantic provinces of Canada, vessel quota programs have also been put in place.[7] As on the Pacific coast, they have grown "organically" out of a regulatory regime aimed at safeguarding the fish stocks but otherwise fairly indifferent toward the efficiency of the industry. In some groundfish fisheries in Nova Scotia catch quotas have been assigned to communities. Some of these communities have opted for individual quotas.

Collective Action by the Fishing Industry

In one way the initiatives from the fishing industry for establishing exclusive use rights to fish are more difficult than the 200-mile issue was for the coastal states. The 200 miles amount to territorial rights for a single entity (a state). In most circumstances such rights would be all but useless for individuals in the industry; the fish are simply too mobile for territorial use rights to make much sense for individuals, and usually also for a group of individuals living in a certain community. Exceptions are sedentary or relatively stationary species. Territorial use rights have been claimed for such species and with some success. The lobstermen of Maine are a famous example.[8] In the Pacific there existed in earlier times, and in some places apparently still exist, territorial use rights which may have been of use for protecting semi-stationary fish stocks from overexploitation.[9] The territorial use rights in Japan were mentioned in an earlier chapter. But for cod or herring (which migrate hundreds of miles), let alone tuna (which traverse the great oceans), such rights would be useless. In fact, some states were fully aware of the lack of instrumentality of the 200-mile limit in some cases and defined their negotiating position at the Law of the Sea Conference accordingly. The United States sought functional arrangements which defined rights to stocks as such. It largely succeeded with Pacific salmon, the fishing of which is prohibited on the high seas, making it exclusively American (or Canadian), but failed over tuna.

Given that territorial use rights will not do, fishermen have to deal with the fish resources as being shared among all of them (and perhaps with fishermen of other countries as well), yet somehow define exclusive rights that have value for them individually. This pretty much leaves them with either fishing licenses or individual quotas as useful instruments. When a fish stock is entirely contained within the territory of the state they live in they might perhaps try to go for ownership of the stock, but given the sometimes traumatic history of establishing the much less ambitious ITQs this would not be easy. Given the widespread use of a limit on the total catch as a management tool, it is not surprising that they have mainly chosen individual quotas.

But before they get that far, fishermen have to overcome a number of obstacles that stand in the way of acting collectively. One set of obstacles pertains to the allocation of quotas. This is a zero-sum game: with a given total catch, if Fred gets more, Dave will have to get less. Different criteria will change the amounts Fred and Dave receive. Much controversy has occurred over what rule should determine the allocation of quotas. The rule that appears most fair, and in fact the one that has been most widely used, is catch history. The rationale behind this is simply that the new regime should not make anyone worse off. A rule based on catch history would seem to do just that. If no quota regime were put in place, it is not unlikely that a particular boatowner would take the same share of the total catch as he did previously. And if the fish stocks recover and the total catch increases, he would share in the gain with everyone else in proportion to his share of the fishery.

Neat as this may sound, it does not fully take care of all aspects of fairness one may identify. One is that boatowners come and go. Those who have just invested in a new and expensive boat would not be well served by only taking into account catch history. When the industry has taken the initiative to implement a quota regime, or been consulted about how quotas should be allocated on the first round, it has often come up with compromise formulas that have given some weight to factors reflecting investment rather than catch history. In the Southern bluefin tuna fishery in Australia, the amount of capital invested was accounted for together with catch history.[10] In the groundfish trawl fisheries and the sablefish fishery in British Columbia, vessel length was given a weight of 30 percent and catch history 70 percent in a quota allocation formula. In the surf clam fishery off the eastern seaboard of the United States, vessel size was given a weight of 20 percent. Iceland experimented for a time with "effort quotas," to allow people to establish "catch history," but with deplorable economic consequences.

Often, however, antagonistic interests within the industry appear to stand in the way of a collective action for a common good. Different parts of the industry may feel little commonality with one another, because of different technologies and business cultures. Small boats are typically owned by just a few persons who know each other well, or maybe a single person or a family, large trawlers or purse seiners by corporations which often own many vessels and are vertically integrated, processing the fish and selling it under their own label. Small-scale operations typically can count on greater sympathy from the public and politicians than large-scale operations. Everyone loves the family farm but curses the large corporation, even if our material well being is due in no small measure to the fact that the former is largely gone out of business while the latter provides us with energy, communications and even food.

Antagonism between different kinds of operations is not new, and what was once novel and threatening becomes traditional and not so dominant after a while. In the 1700s long lines (i.e., lines with many hooks laid horizontally over long stretches and pulled in hours later) were introduced in the cod fisheries in Northern Norway. This caused such consternation among those who used the traditional hand line (a line with a sinker at the end and one or just a few hooks and attended constantly) that complaints were repeatedly made to the authorities about this new and threatening method of fishing. It was alleged to be affordable only by rich people and to encourage fishermen to be foul-mouthed. The protests finally reached the King in Copenhagen who ruled that only the hand lines, known since time immemorial, should be used.[11] But in the end the long line prevailed. In more recent times, the authorities in Norway have followed up this tradition, for example by banning the use of purse seines in the Lofoten fishery in the 1950s, seen as threatening for, *inter alia*, fishing with traditional gear such as long lines.

Use Rights: Property or Not?

In the cases where governments have accommodated the initiatives from the industry toward establishing exclusive use rights, and even in cases where governments themselves have taken the initiative, they have been extremely reluctant to recognize such rights as full-fledged property rights, guaranteed by the law of the land in the same way as property rights in general are guaranteed and revocable only with adequate compensation. The American ITQ programs explicitly contain a clause stating that these are not rights but "privileges" revocable at any time without compensation. The Canadian regulations use a similar language. The Icelandic parliament has debated at some length whether or not the ITQs amount to property rights, and at one point at any rate the majority seemed to think that they are not, concluding that ITQs could not be mortgaged.[12] This is ill advised, as it only serves the purpose of making it more difficult for people without capital of their own to get into the industry. It is also contradictory, as many of the same legislators who oppose allowing ITQs to be mortgaged have expressed concerns over the difficulty for such people to enter the industry. The Icelandic Supreme Court has stated that ITQs do not amount to property rights in perpetuity, but they have nevertheless been treated as property in marital dissolution cases. Only the New Zealand government has gone the whole way and defined the ITQs as property held in perpetuity, unless transferred to someone else.

Some legal experts in Norway have maintained that allocating rights to public resources in perpetuity would be unconstitutional, since parliament cannot make binding obligations on behalf of its future members. Formally there is some truth in this, but on the other hand even the constitution can be changed, and ordinary property rights which many constitutions explicitly protect can be and have been confiscated in the public interest. Revolutions have been made over the issue of property rights, against which written constitutions have been of little help. The point is that private use rights to fish or whatever public resource it is can and will be upheld as long as there is sufficient political support among the public at large, and if that support no longer exists, even the most entrenched property rights will be of no value.

The reluctance to recognize ITQs as full-fledged property rights serves little purpose other than detract from their instrumentality. The reasons probably are ideology and unwillingness of governments to relinquish ultimate control. Many people abhor the idea of turning a public resource into private property and making it possible for private individuals to enrich themselves by using such resources for their own gain. Those who hold such opinions apparently fail to realize that what has made nations wealthy is private property rights and strife for private gain, together with a system of governance which has distributed the fruits of this reasonably widely. Or maybe they lack the ability of applying this insight in a new and unfamiliar area. The attitude to modes of organization, including what is common and what is private property, is shaped by tradition and history. The difficulties of establishing private property rights to fisheries in countries whose economic system and national wealth is built on property rights and market processes is in many ways parallel to the difficulties in the former Soviet Union to establish private property rights to land. Not only was the concept alien, but generations of Soviet citizens had been brought up in the belief that private ownership of means of production, including land, was a barbarian relic. Reports from the former Soviet Union in the mid 1990s told of workers on collective farms driving their cattle to their pastures of old even if the land was supposed to have been sold to private interests, and of barns on privatized land being burnt down in anger by opponents of privatization.

The reluctance to recognize use rights in fisheries as private property detracts from their instrumentality by creating uncertainty about their validity in the long term. This favors short-term gains over long-term ones, shortens the time horizon in any investment analysis (or raises the discount rate applied), and makes rights holders less interested in preserving the productivity of the fish stocks than they would otherwise be. It also makes it more difficult for new entrants to buy their way into the fishery. When market values of quotas or licenses are high, new entrants normally have to borrow money to enter the fishery, just as they have to borrow money to buy a boat. But while a boat can be mortgaged as a security for a loan a fish quota cannot, unless it is recognized as a private property. Hernando de Soto would argue that the lack of recognition of fish quotas as private property prevents them from being converted into capital and used for expanding the operations of whoever happens to hold them.[13]

It is possible, however, that these shortcomings of fish quotas as private property are more apparent than real, at any rate in certain settings. Ways can sometimes be found around the problem of not being able to mortgage fish quotas. In the surf clam and ocean quahog fisheries off the eastern seaboard of the United States, the lending institutions assume the ownership of the quotas until the loan has been paid. This, however, is only possible when the ownership is not restricted to individuals who actually do the fishing, as the case is in the Alaska halibut fishery. In Iceland, before quotas could be mortgaged, they could not be sold unless the institutions having lent money to the prospective seller had been notified and given their consent. The revocable tenure may also be less of a threat than it looks at first sight. There appears to be only one case (the Faeroe Islands) where an ITQ system has been revoked once it had been put in place, but modifications and tinkering at the edges are not uncommon. The expectation seems to be that, once in place, such systems are there to stay, and fishermen seem to make long-term decisions accordingly. After such systems have been put in place the industry has begun to take a much stronger interest in the long-term productivity of the resources, often financing both stock assessments and biological research. The pressure for higher quotas in the immediate future has abated or has even been reversed in favor of rebuilding stocks.

It is highly likely that governments will be reluctant to exercise their option to revoke ITQs or fishing licenses, even in the cases when they have explicitly reserved that option for themselves. The strong economic interests attached to tradable quotas with a high market value will ensure that there would be intense lobbying and political pressure against, and in the wake of, such a move. There would be much upheaval in the industry, with new entry and all the problems of overfishing this would entail in case the industry were opened up. Quota owners might file damage suits and possibly win if stripped of a valuable asset purchased in the faith that the system would continue.

Formal recognition of quotas as a full-fledged property may be less important than it looks at first glance. This did not stop the government of New Zealand from changing the quota rights from a fixed number of tonnes to shares in the total allowable catch, but it is likely to have given strong support for the industry's claims for compensation, which ultimately led to the abolition of the resource rentals. What the argument boils down to is the willingness of society to support a particular institution. Tacit support of an informal arrangement can be worth more than laws on the books. Written laws and constitutions can be changed, and are changed when the stipulated majority perceives a need for doing so. Customs that are commonly accepted without being codified in written documents may have a much stronger hold.

The presence of ITQs has generated a certain amount of discussion within the legal profession about whether or not they are "property," and if they are, of what kind. Some of this discussion has taken place in connection with rulings in disputes involving economic interests tied to ITQs. Given the novelty of ITQs, it is perhaps not surprising that the views within the legal profession differ. Some differences of opinion are undoubtedly due to the pertinent laws and regulations being different in different countries, but even within one and same country legal experts do not always agree on what kind of an animal this is. But if it walks like a duck and quacks like a duck, what could it possibly be? Australian courts seem to agree that ITQs are a form of property but disagree on their exact nature.[14] Courts in the United States have considered several cases involving disputes over ITQs. In the *Foss* case, the Ninth Circuit Court found that "there is no doubt that the IFQ permit is property. It is subject to sale, transfer, lease, inheritance, and division as marital property in a dissolution."[15] A different court, in a different case (*The Sea Watch International et al. v. Mosbacher*) saw the quotas merely as regulatory instruments. In dismissing the plaintiff's argument that the quotas in the surf clam fishery involved privatization of a common resource, the court stated: ". . . the new quotas do not become permanent possessions of those who hold them, any more than landings rights at slot-constrained airports become the property of the airlines, or radio frequencies become the property of broadcasters. These interests remain subject to the control of the federal government which, in the exercise of its regulatory authority, can alter and revise such schemes, just as the Council and the Secretary have done in this instance."[16]

Rather than being irreconcilable opinions, these differences reflect the multi-dimensionality of property rights. They are seldom if ever absolute, they function within constraints that vary from one kind of object to another and over time. Some of these constraints may be unwritten and implicit rather than written and explicit. Again this reflects the fact that property rights are embedded in society and must be supported by a critical mass of those who hold power. And society can, similarly, restrict property rights. Residential housing is sometimes subject to rent control. The height of buildings to be erected on a given piece of land may be limited, particularly if it would spoil the view of somebody in the neighborhood. Buildings deemed of historical value cannot be modified at will. Owners of land may have to put up with rights to innocent passage, and even picnicking and berry picking on their land. In many countries mineral rights typically belong to the state even if the surface is individually owned. And so on and so forth.

Community versus Individual Rights

Above we have mostly been discussing individual rights, i.e., fish quotas or fishing concessions being held by individuals or firms. Use rights could also be held in common by some group of people, circumscribed, for example, by requiring residence in a certain area. An example of this is the Community Development Quotas in the Alaska pollock fishery in the United States. Ten percent of the total catch quota in this fishery is assigned to small, isolated communities in western Alaska. As the name indicates, the purpose is to promote economic development in these communities rather than exploit any comparative management advantage these communities have; in fact they have little or no tradition in this fishery. Most of these communities have elected to cooperate with established firms in the pollock fishery about fishing these quotas, receiving monetary payments in return or job opportunities for some members of the community.

The community development quotas in the Alaska pollock fishery are not a good example of the types of communal rights being discussed in a large and growing body of literature on common property. This literature, penned mainly by anthropologists and social scientists other than economists, is concerned with self-governing arrangements in communities which have traditionally depended on the resource under consideration. Examples would be fishing communities with a long history (the communities in western Alaska never fished for Alaska pollock, and this fishery is in fact quite new). There is a curious undercurrent of Panglossian belief in traditional wisdom running through much of this literature. It typically takes exception to Garrett Hardin's fable of the tragedy of the commons, arguing that those who shared the commons surely must have found a way to avoid it, being limited in number and realizing what would be in their common interest. There are indeed examples of successful management of common resources such as irrigation systems and pastures. Some of these are documented in Elinor Ostrom's pioneering and much quoted book.[17] What is noteworthy, however, is that in all the examples of success it has been possible to restrict the number of users of the common resource in some way or to count on a natural limitation of users (the use of irrigation systems is limited by whose land they traverse). This is very different from open-access fisheries where the number of users is essentially indeterminate.

For some people, apparently, communal access rights appear more ideologically appealing than private access rights. It is difficult, however, to see why those who share Rousseau's dislike for the impostor who says "this is mine" should be so much more enamored of property rights held in common than private rights. The difference between "this is mine" and "this is ours" need not be terribly great; if the latter has any meaning it means that those who are not among "us" are excluded. The distinction between "us" and "them" in community-based rights systems can be fairly arbitrary and seemingly unrelated to any principles of fairness. Access to one of the Alpine pastures of Switzerland is limited to those who carry family names that have been in the village since the 1500s or longer.[18] Community-based access rights to fish are often and perhaps typically based on belonging to a caste or a religious or ethnic group, or long residence in a certain place. In a community in Sri Lanka, fishermen from "traditional fishing families" can have two motorized craft for catching shrimp, but others only one. Membership in the Fishermen's Association in one Japanese village is restricted to families who have lived in the village for at least one generation.[19] Such rules may be very effective, as long as they go unchallenged, but it is difficult to see why they should be more appealing than rights being bought and sold in a market, or perhaps tendered by governments as the ultimate rights holders. It is also difficult to see why rule enforcement within exclusive and self-constituted groups should be more appealing than law enforcement through the normal, civilized procedures of the modern state. Justice within self-constituted groups may be effective but can be fairly rough, and the rougher it is the more effective it becomes.

5   Successes and Failures: New Zealand, Chile, Norway, and Canada

The development of exclusive fishing rights in these four countries has little in common except that there has been a strong, and in the long run probably irresistible, undercurrent toward establishing such rights due to the ever increasing pressure on the resources. In all four it has turned out to be necessary, and increasingly so, to limit the catch from the most coveted and easily accessible fish stocks and, consequently, to limit the access to these resources. There has been a growing appreciation in the industry of the need to keep newcomers out, in order to protect vested interests. But the approach taken to establish exclusive use rights, and the success or failure this has met with, has varied from place to place, underlining the importance of history, traditions, and other institutional circumstances specific to each country. New Zealand established individual transferable quotas early, swiftly, and successfully, although not without problems. Chile failed, and so did Norway, despite historical and institutional differences. In Canada the story has been mixed, having in part to do with the different circumstances on the Atlantic versus the Pacific coast..

New Zealand and Chile are both known for wide-ranging economic reforms in the neoliberal spirit, but there the similarity ends; these reforms were designed and implemented under quite different circumstances. In New Zealand the reforms were carried out by a new, democratically elected government. Somewhat surprisingly, at any rate for a person from the northern hemisphere, these reforms were carried through by the Labor Party of New Zealand, which in 1984 replaced the conservative National Party. In Chile the neoliberal reforms were carried out by the military government led by General Augusto Pinochet, who sought advice on economic policy from a group of economists educated at the University of Chicago, sometimes referred to as the "Chicago boys." Despite all the differences, there was one commonality between the governments of New Zealand and Chile in the late 1980s; both could carry out their reforms without much political opposition. At that time New Zealand practiced the British first-past-the-post system, electing only one person in each constituency, which gave a single party firm control over the legislature. A determined government could therefore push through its reforms without running into much of an opposition, in parliament at any rate. There is a parallel here with the radical market-oriented reforms of Margaret Thatcher in Britain, which took place at about the same time. Proportional representation with its coalition governments is good for consensual approaches but less so for radical reforms that break with the past.

The governments of Chile and New Zealand took initiatives to implement ITQs in their fisheries, a method which relies on market forces and so accords with neoliberal thought. But while New Zealand succeeded and has become something of a role model for other countries seeking to take this approach, the Chilean approach failed. Industry was the decisive factor; while the New Zealand reform had the support of at least a critical mass of the industry the Chilean reform was defeated by groups within industry.

Norway, by contrast, has not made itself known as a paragon of neoliberal economic experiments. That notwithstanding, the Labor government of Gro Harlem Brundtland (1986–1996) did introduce some decisive economic measures in the neoliberal spirit. The capital market was deregulated and the determination of the interest rate removed from the political sphere and vested in the central bank. Endemic inflation was weeded out by fixing the exchange rate of the national currency relative to the European Currency Unit, the precursor of the Euro. This was a harsh and costly reform; unemployment grew to heights not seen since the Great Depression, and several banks either went broke or lost all their equity capital and had to be bailed out. An attempt was made to introduce ITQs, but this proved more difficult than reforming monetary policy. One reason for this failure was the view, well entrenched among local politicians, the bureaucracy, in academia, and even some industry circles, that the role of the fishing industry is to provide employment and maintain communities in areas where there are few other opportunities. With an outlook like that, efficiency enhancing mechanisms like ITQs have little role to play and would even run counter to the objectives of maintaining employment and supporting disadvantaged communities.

The fisheries of Atlantic Canada have a certain similarity with the fisheries of Norway in this respect. Atlantic Canada, particularly Newfoundland, depends heavily on fishing, just like some coastal communities in Norway. As in Norway, the fisheries of Canada have not been seen exclusively, or even predominantly, as contributing to the nation's wealth. On the Atlantic side, especially in Newfoundland, the fisheries have been seen as a provider of employment, without much attention being paid to whether it is particularly gainful, from an overall economic point of view. The industry has received substantial subsidies, in particular through an unemployment insurance scheme which has been particularly generous for fishermen. By fishing for a few weeks, fishermen have become entitled to unemployment support for the rest of the year. This has undoubtedly kept up the employment level in the fishery and blunted the incentives people would otherwise have had to move to other parts of Canada to find better opportunities.

The fisheries of Canada are the prerogative of the federal government, and so fisheries policy is framed at the federal level and not at the provincial level. With the fisheries being seen as a part of social policy rather than a wealth-generating activity, it is perhaps not surprising that the Canadian government does not have a clear policy encouraging fisheries management regimes based on private use rights. As far as use rights are concerned, the attitude of the federal government is perhaps best characterized as readiness to tolerate such regimes if the industry asks for it. In the last part of this chapter we take a look at the development of ITQs in three fisheries off the Pacific coast of Canada. This story is quite interesting and in some respects contradictory to received wisdom.

New Zealand

New Zealand is one of the countries that benefited most from the 200-mile zone. Its location as an island in the South Pacific far from other countries ensured that it got one of t' e largest zones in the world. In the 1960s and early 70s distant water flee s from the Soviet Union, Japan and Korea fished in the waters that in 1977 became the exclusive economic zone of New

Zealand. Fishing fleets from these countries discovered the orange roughy, a fish that was to become one of the most valuable resources for the fishing industry of New Zealand.

After the 200-mile zone was established, the foreign fleets were expelled, and New Zealand developed its own deep-sea fishing fleet. Only a handful of companies were engaged in this fishery. This was a greenfield situation with only a few, large players, and the New Zealanders, mindful of what had happened to open-access fisheries all over the world, decided not to repeat it. In 1983 they introduced individual vessel quotas in their deepwater fisheries.

Soon afterward it was decided to use ITQs not only in the deepwater fisheries but also in the traditional fisheries of New Zealand. The preparations for the system were well under way in 1984 but the whole process of quota allocations took some time; the system did not take effect until 1986. As the ITQs were introduced, the catches from the fish stocks involved were reduced by a proportional reduction in ITQs relative to previous catches. The introduction of the ITQs in the traditional fisheries of New Zealand thus were a part of a process of trying to rebuild fish stocks, a bit like in Iceland, to be discussed in the next chapter. No less important, the ITQs were also a part of a general economic reform taking place in New Zealand at that time under the Labor Party Prime Minister David Lange. Immediately after World War II, New Zealand had been at the top of the OECD league in terms of GDP per capita. By the early 1980s it had fallen to a middle position, a decline widely ascribed to inadequate economic policies. New Zealand had high tariffs sheltering industries producing for its tiny home market (at that time the population of New Zealand was about 3 million), and many sectors of the economy, including its highly competitive agricultural sector, were heavily subsidized. Making the economy more competitive and open to trade through lowering tariff barriers and dismantling subsidies was seen as a better way forward that hopefully would restore New Zealand to its pride of place. ITQ-managed fisheries fit excellently into this picture, the main purpose of ITQs being to make fisheries as economically efficient as possible.

The groups behind the ITQ program consisted of leading politicians at the time, government officials, and parts of industry. The traditional fishing industry of the country appears to have supported the program without much opposition. It is noteworthy, however, that part-time fishermen had been excluded from the fishery at the time ITQs came to be debated. A new fisheries law in 1983 introduced a permit scheme, by which over 2000 part-timers, almost one-half of all commercial fishermen, were removed from the industry.[1] This is likely to have helped promote later support by the industry of the ITQ system. The exclusion of the part-timers hit some Maori fishermen and probably contributed to the dispute that emerged between the Maori and the government over the ITQ system.[2]

The industry's support of the ITQ system, which was strong but not unanimous, can be explained by the gains fishermen could expect to make. The stock recovery engineered through the quota buy-back could be expected to benefit the industry. The ITQs would prevent the erosion of profits in the industry that otherwise would take place through competing for the largest possible share of the catch. The quota rights were issued as fixed tonnages in perpetuity, giving the industry a guaranteed annual catch of fish, either as real fish or paper fish. The idea was that when the stocks were in a bad shape and would not support the sum total of quotas issued to the industry, the government would buy back quotas as needed to keep the total catch within the appropriate limits, and sell additional quotas in better times. The government thus took all the risk associated with fluctuations in the fish stocks. It is possible to design a system like this in such a way that the government makes money on its transactions in the long term, making it in effect a resource rent recovery scheme.[3] This may have been the intention, but things did not turn out that way, as will be discussed below. The quotas were given to the industry free of charge but a resource rental was planned, and in fact put into effect in the orange roughy fishery.

It soon turned out that the government had made a major mistake in its assessment of the orange roughy stocks. This assessment had been based on the early development of the fishery while it was still fishing down virgin stocks, but a more sober assessment showed that the stocks would not be able to sustain the fixed tonnage quotas that had been given to the industry. Keeping the system as it was would therefore have meant that the government would have had to buy quotas of orange roughy from the industry every year, in effect subsidizing the industry. In 1990 the quotas were changed to share quotas, i.e., each quota holder gets a certain share of the total catch permitted every year. This in effect transfers the entire risk associated with variations in the fish stocks to the industry.

Unsurprisingly, the industry did not accept this without a fight, arguing that the quota allocations had initially been advertised as secure property rights in perpetuity to given amounts of fish. In the end the industry accepted the change but the government waived its resource rentals, opting instead for a cost recovery program.

This notwithstanding, the ITQs in New Zealand are among the most firm property rights one finds in capture ocean fisheries anywhere. They are explicitly perpetual, they can be transferred almost without restrictions (there is a limit on how much quota can be held by any one firm, and foreign interests cannot own ITQs), and they can be mortgaged. In New Zealand there appears to have been little resistance to so defining property rights to a public resource, unlike in many other countries. But even here the rights are use rights, not rights to the resources themselves. It is still the New Zealand government that decides how much can be caught from each particular stock in each particular year, even if the industry plays a major and an increasing role in this.

Even if the New Zealand system has been widely hailed as a successful management system, more firmly based on individual property rights than anywhere else, it has over the years been subject to several modifications, prompting one commentator to refer to it as "unfinished business."[4] These modifications have less to do with inherent faults in the quota management system than with other developments, which it was never designed to cope with, so the assessment that "the basic structure of the system has remained intact" appears more to the point.[5] An exception is the aforementioned mistake of setting quotas as fixed tonnages based on overoptimistic assessments of what the fish stocks could yield. Two developments are particularly important in this regard, the conflict with the indigenous Maori population and the environmental concerns that have increasingly come to the fore.

The conflict with the Maori population stems from the Treaty of Waitangi. In this treaty, concluded with the Maori chiefs in 1840, the then colonial power, Great Britain, pledged to respect the traditional fisheries rights of the Maori. The ITQ program violated these rights, in the view of the Maori chiefs. To an outsider the case has many of the hallmarks of rent seeking with vicarious arguments. The dispute was finally settled by allocating a certain amount of the annual total catch quota for certain stocks of fish to the Maori and by buying up a share of the largest New Zealand fishing company, Sealord, and giving it to the Maori. Sealord bases much of its activities on deep-sea fishing for orange roughy, which is anything but a typical Maori traditional fishery. Even if the Maori, like other Polynesians, were the greatest seafarers the world has ever seen, their primitive technology did not allow deep-sea fishing.

Environmentalists have become increasingly engaged in fisheries issues over the years, in New Zealand and elsewhere. Their crusades against seal and whale hunting are famous, or notorious, depending on how one looks at it. Their arguments derive less from any concerns about harming the yields from these stocks than from sanctity of certain forms of marine life, be it super intelligent whales or cuddly seals (cuddly, that is, when stuffed, as the specimen the famous actress Brigitte Bardot held in her arms in a widely published photograph). Similar ideas are becoming increasingly current in discussions on fisheries management. Fish stocks are to be preserved as such for the protection of biodiversity, not for the purpose of any material utility, much as stocks of wild animals, previously hunted for food or as threats to domesticated animals or crops, have become protected wildlife. Greenpeace has sued the Ministry of Fisheries for setting a higher overall catch limits for some stocks than warranted in order to rebuild them to the maximum sustainable yield level, which by law the ministry is supposed to do.[6]

It goes without saying that a quota management system is ill suited to deal with questions of preservation for purposes other than material benefits. The quota management system is primarily about obtaining maximum economic benefit from whatever quantity of fish one is permitted to take out of the sea. This does not conflict with any conservationist goals; these can be taken care of by setting the total catch quota as conservatively as such goals would demand. But if the quota management system is to be developed to the point that the industry itself sets the total catch quota, it is understandable that environmentalists become alarmed. The degree of conservation one can hope for on the part of industry is the one that maximizes its profits in a long-term perspective. It is quite possible, and perhaps even likely, that this would result in a degree of conservation adequate from a broader point of view; in order to maximize profits in a long-term perspective the fish stock has to be maintained at an appropriate level. One advantage attributed to the quota management system is fostering industry interest in the long-term productivity of the fish stocks.

In the words of a previous minister of fisheries, ". . . the protection of the balance sheet value of the property right, the sense of ownership and commitment to the fishery, have led us largely from an industry of hunter gatherers seeking to beat each other for the last fish to an industry of seagoing farmers."[7] But there are exceptions, cases where the natural growth of fish stocks is slower than the return required by owners of capital, making any deferment of catches a poor investment and pointing toward "mining" such stocks to extinction rather than preserving them for sustainable profits. The slow growing orange roughy could be one such case.

At the moment it is uncertain in which direction the quota management system of New Zealand will develop. Will it become increasingly the responsibility of the industry itself, or will the focus of industry be considered much too narrow, partly because of environmental concerns and partly because of a conflict with recreational interests with intangible and non-commercial benefits? Certain management tasks have already been delegated to the industry. The industry contributes toward financing the management costs and has, therefore, demanded a greater say in how the money is used. This has not always been a happy relationship; the industry has frequently complained of being overcharged, and at least on one occasion the New Zealand Parliament has lowered the contribution demanded from the industry. It has been alleged that the industry is mainly preoccupied with spending less on the tasks that it is being charged for and more on those that are being paid for with public money, but less concerned with increasing the overall efficiency of management, contrary to views often expressed that increased influence of the industry would enhance the efficiency of management.[8] That notwithstanding, the industry plays a very active role in the assessment of fish stocks; it hires its own stock assessment consultants who check on what the experts hired by the government are doing and provide their own assessment; in fact, stock assessment in New Zealand is probably best described as a joint effort by industry and government. Industry takes this very seriously; it has for a number of years hired specialists of world renown from overseas and had them spend months in New Zealand on this work. But one could also envisage the reversal of this trend, due to an increasing emphasis on public good aspects of fisheries, be it because of environmental concerns or recreational interests.

What has the New Zealand quota management system accomplished? We would expect it to result in a greater economic efficiency, in the form of less capital invested in fishing boats and fewer fishermen than there would be under open access. It may seem surprising, therefore, that both the number of boats and the number of people employed in the catching sector increased over the period 1987–1995.[9] But this is not the whole story. Over the same period the total catch in the New Zealand fisheries has more than doubled, which is likely to have required more capital and manpower.[10] With few exceptions, the catches from the most important fish stocks around New Zealand have increased handsomely since the quota management system was put in place. Figure 5.1 shows the catches from the ten most important fish stocks around New Zealand in the period 1980–2001. The catches taken from seven of these have trended upward with minor interruptions since 1986. The catches from two stocks (arrow squid and southern blue whiting) show major fluctuations but an upward trend. Only the orange roughy has trended downward, due to the initial depletion of a pristine stock, as already discussed. Since the said increases in catches have continued over a long period, it is difficult to avoid the conclusion that the management of these fisheries has been a success in terms of conservation. Strictly speaking this is not necessarily due to the ITQ system; it is due to setting the limit on the total catch appropriately, but indirectly some credit is probably due to the quota management system; the industry has considerable influence on the setting of the total catch limit, and the quota management system is indeed likely to have provided incentives for the industry to argue for setting the total quota conservatively. A further sign of success of the New Zealand fisheries management system is that the hoki fishery, the most important fishery in volume and in value, was the first one in the world to get certification from the Marine Stewardship Council.

There are other, clear indications that the ITQ system has increased the economic efficiency in the fisheries. Such efficiency gains would be reflected in a high and rising price of quotas—high because economic efficiency leads to higher rents which become capitalized as market prices of quotas, rising because the introduction of the system would be expected to gradually eliminate inefficiencies. This is exactly what has been found in a recent study of price formation for fish quotas in New Zealand.[11] After

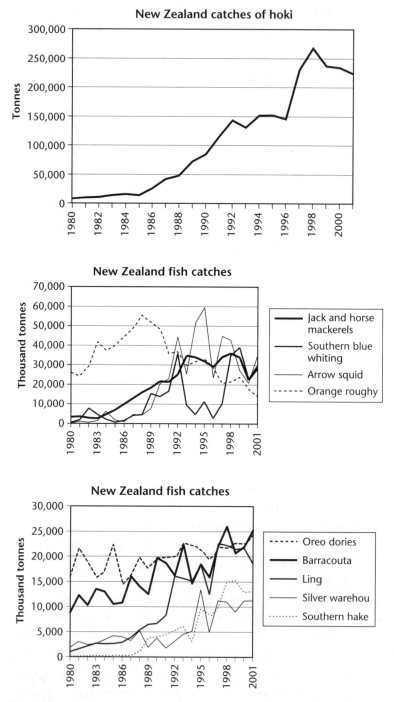

Figure 5.1
Quantities of the ten most important fish species in New Zealand. Source: fishery statistical database, UN Food and Agricultural Organization.

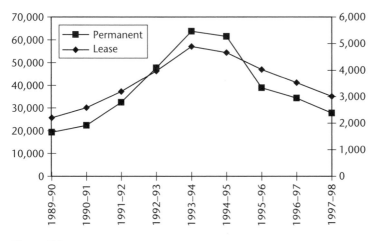

Figure 5.2
Price of snapper quotas in Area 1 in New Zealand, permanent and lease (within a year), in constant 1998 NZ dollars. Source: Batstone and Sharp 2003.

correcting for a host of factors, it was found that the quota prices have increased over time since the ITQ system was put in place in 1986. As an illustration, a time series of quota prices for one particular species and management area is included (figure 5.2). The prices fluctuate over time, due among other things to changes in export prices of fish and the general conditions in the New Zealand economy, but with an upward trend.

Chile[12]

In the late 1980s the military government in Chile tried to implement an ITQ-based management system. The "Chicago boys" were, however, late converts to the ITQ gospel. In its early days the military government had privatized the Chilean fishing industry, seeing it as a priority to encourage free enterprise and competition in the fishing industry instead of reigning it in by unnecessary regulations. From a somewhat restrictive policy for issuing fishing licenses the regime was changed in 1978 to open access where fishing licenses were a mere formality. At that time the fisheries in southern Chile were still at a developing stage. In the north the anchovy and the sardine had long been exploited intensively, and signs of overexploitation were soon to emerge.

In the 1980s it became increasingly clear that the limits of what the fish stocks could yield were being approached, and possibly attained or even exceeded in the north. The economic advisors of the government realized that some regulation of the industry was necessary. True to their prioritization of economic efficiency they went for ITQs. These ideas were not well received in the industry. The decisive opposition probably came from the Angelini group, which dominated the anchovy and sardine fisheries in the north. There was also opposition from the fishworkers' union, both for ideological reasons (privatization of a common resource) and because of fear that catches would be restricted and jobs would be lost.

Several factors accounted for the opposition by the Angelini group. In the 1980s the government had occasionally set quotas for the total catch of the most heavily exploited stocks (horse mackerel, sardines and anchovies). These limits were guidelines rather than effective constraints; they were never enforced and routinely exceeded; sometimes the catch was more than twice as high as the quotas would have allowed. The industry apparently did not see its interests being well served by any government imposed catch limits. Whether or not fishing in excess of the recommended quotas has had an adverse effect on catches is, however, an open question. Figure 5.3 shows the development of the catches of anchovy and sardine in Chile. The catches of sardine peaked sharply in the mid 1980s and then went into a decline and have now almost disappeared. The catches of anchovy fluctuate enormously from year to year, but if anything with a slightly rising trend. The industry argues that the decline of the sardine is a part of an environmentally determined cycle by which sardine is replaced by anchovy or vice versa, and that it does not much matter how the productivity of the ocean is exploited (both sardines and anchovy are processed into meal and oil). Figure 5.3 certainly indicates that this might indeed be the case. The sardine and anchovy stocks in Chilean waters are affected by the El Niño, as the stocks off Peru, and Peru and Chile in fact share these resources to some extent.

The second reason why the Angelini group opposed the ITQs was that it wanted to get into the southern fisheries but had little catch history to count on there. The ITQs were to be distributed on the basis of recent catch history, which would have suited the Angelini group just fine in the north where it dominated the industry; it used to take more than one-half of the total catch in the industrial fisheries in the north.[13] To get into the south-

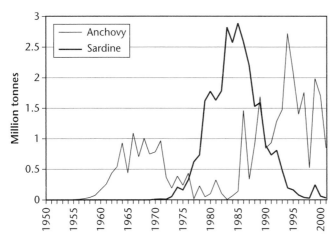

Figure 5.3
Catches of anchovy and sardine in Chile. Source: fishery statistical database, UN Food and Agricultural Organization.

ern fisheries, the Angelini group would have had to buy its way in under an ITQ system, whereas it could get in virtually for free under the existing system. The Angelini group was at the time planning to divert some of its boats into the southern fisheries, because of falling catches in the north.

A new fisheries law (the Merino Law, named after the Minister of Fisheries in charge at the time) establishing ITQs was to take effect in March 1990 but encountered problems under the new, democratically elected government. This political transition is likely to have contributed to the later revision of the proposed law and undermined the ITQ regime. Even if the military government resurrected the Chilean economy and put it on a more promising track than found in other Latin American countries, its legacy was in many ways troubling. While later governments have largely kept intact the economic structure inherited from the Pinochet years the fisheries law was substantially revised. The lobbying efforts of the Angelini group and others bore fruit. For fully exploited fisheries the quota scheme was made unworkable; only half of the total catch quota can be put under ITQs, and ITQs can only be implemented gradually by putting 5 percent of the total catch under ITQs by auction each year, which would make for a ten year transition. So far this has not happened in any fully exploited fishery. For new fisheries ITQs can be allocated by auction. This has been done in four small but valuable fisheries; two deep-sea fisheries in the south (orange roughy and Patagonian toothfish) and two crustacean (red and yellow prawn) fisheries. This quota system is similar to the one practiced in Estonia for a few years early this century in that 10 percent of the quota "withers away" every year and has to be bought back if the full quota is to be retained.

The distinction between the use of ITQs in fully exploited versus new fisheries is difficult to understand other than as a consequence of lobbying by parties already established in the fully exploited fisheries, as argued by Peña-Torres (1995). Why should only 50 percent of the total catch in fully exploited fisheries be under ITQs while 100 percent is permissible in new fisheries? The fact that no one has opted for ITQs in fully exploited fisheries speaks for itself.

In the meantime, overexploitation has developed and gotten worse in the most important Chilean fisheries. An opportunity to prevent this and to take advantage of a greenfield situation was missed. The Chilean government is now trying to get to grips with the situation, and ITQs are again a much-talked-about option. An attempt in the late 1990s to allow the use of ITQs, primarily in the fishery for horse mackerel, troubled by the El Niño event, was defeated by a somewhat unholy alliance of large-scale operators and small-scale fishermen. Nevertheless, the country has already taken the first steps to apply individual quotas on a major scale. A new law valid for 2000–2002 authorized enterprise quotas in fully exploited fisheries, by which different firms were allowed to enter into cooperative arrangements and coordinate their fishing operations, a bit similar to what has happened recently in the Pacific groundfish fisheries in the United States. This has led to some boats being laid up, so lowering the costs of fishing. A side effect of this has been that some crew have been put out of work, to which some have reacted with vocal protests and semi-violent actions.[14] As it was due to expire, the law was prolonged a further 10 years.

Norway

Fishing rights that can be bought and sold in fact have a considerable history in Norway; they can be traced back to what happened after the collapse of the herring stocks in the northeast Atlantic and the North Sea. The boats fishing for herring found other stocks to go for but there were too many boats for the fish that was available. To deal with this a system of fishing concessions was introduced for the purse seine fleet which used to catch the herring. The concessions were denominated in volume of cargo capacity (hectoliters). For fish stocks limited by an overall catch quota the quota was portioned out among the vessels according to the capacity they had concession for. The allocation rule varied to begin with according to species and fishing season but soon congealed into a universal rule. The rule identifies a "base quota," which depends on the size of the vessel (figure 5.4). The share of each vessel in the total catch quota is determined by the share of its base quota in the sum total of base quotas. Vessels less than 90 feet or 1,500 hectoliters are not included in the concession system and get a certain allocation in common.

The vessel concessions turned out to be valuable. The reasons were economies of scale and general overcapacity in the industry. After the concession system was put in place it became impossible to acquire new and larger boats unless a concession capacity corresponding to the new boat was removed from the fleet. A boatowner who wanted to replace his boat with a new and bigger one had to acquire a concession for the additional capacity through buying another boat and scrap it or sell it out of the

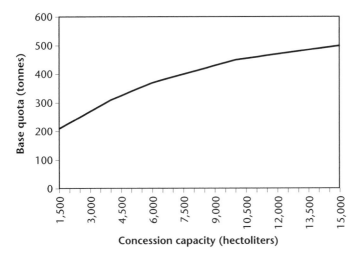

Figure 5.4
Relationship between a Norwegian purse seiner's base quota and its concession capacity. Source: Norwegian Directorate of Fisheries.

country. Even if the rule relating boat quotas to the concession volume discriminated heavily against large boats (figure 5.4), it was still attractive to buy a larger boat in order to acquire additional concession volume. Hence it became possible in Norway to trade quota allocations for the long term, even if it had to take place through the trading of boats.

Originally, however, the government did not intend the concessions to be tradable; this happened more or less by default. The concessions were given to persons owning boats, and they could not be transferred to other boatowners without permission by the authorities. The intention was that concessions could only be passed to next of kin but practice became such that concessions could be transferred to anyone and sometimes even divided up between different buyers, each using his share to get a larger boat or to legitimize the capacity of a larger boat that had already been acquired. One reason why the authorities accepted this practice was that it turned out to facilitate the reduction of the fleet in terms of numbers and to enhance its profitability. In some periods the authorities even encouraged this by providing subsidies to scrapping or selling boats, the concessions of which were then transferred to other boats. It also happened that the government bought boats for decommissioning. The concessions of such boats were nullified, which contributed to a lesser mismatch between the capacity of the fleet and the fish resources available. The need to restructure the fleet was an eleventh-hour insight; in the mid 1970s, a few years before the decommissioning subsidies started, the Norwegian government had subsidized the building of new purse seiners even if it was known that no such vessels were needed and that there was in fact over-capacity in the existing fleet.[15]

The decommissioning subsidies and the withdrawal of concessions led to a considerable reduction in fleet capacity, measured as volume of cargo capacity (figure 5.5).[16] The capacity started to decline in 1979, when the decommissioning scheme was put into effect, and has remained fairly stable since 1989 (most of the decommissioning money was spent before that time). The effect on economic returns was not immediate, however, because of falling catches (figure 5.6), but since the early 1990s there has been a substantial increase in economic returns. Figure 5.7 shows the development in economic returns measured as "wage potential," i.e., what is left after all costs, including capital costs but excluding labor costs, have

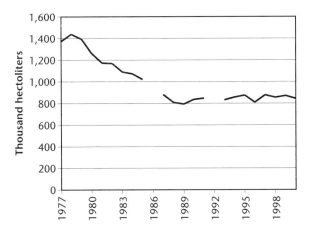

Figure 5.5
A Norwegian purse seiner's total cargo capacity as of December 31 each year. Source:
Statistics Norway, Fisheries Statistics.

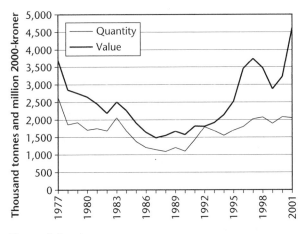

Figure 5.6
Catches of pelagic species in Norway. All values are accounted for in a constant value
of money by use of the consumer price index. Source: Statistics Norway, Fishery
Statistics and consumer price index.

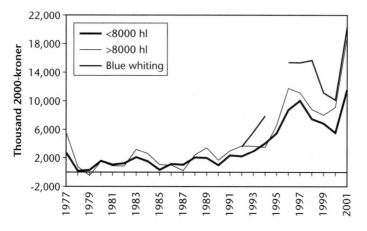

Figure 5.7
Wage potential per vessel for three groups of Norwegian purse seiners; small (less than 8,000 hectoliters cargo capacity), large (greater than 8,000 hectoliters cargo capacity), and vessels with a concession to fish blue whiting. Source: Norwegian Directorate of Fisheries, Lønnsomhetsundersøkelser (cost and earnings studies).

been subtracted from revenues. Without the concession system there is little doubt that the higher economic returns in recent years would have been eroded by the entry of new vessels. Even if the Norwegian concession system is a crude variety of a quota management system, we can take this as an example of how such systems promote efficiency in fishing. This point has not been lost on the authorities; even if their pursuit of economic efficiency has been somewhat half-hearted, the Ministry of Fisheries made the trade in fish quotas more rewarding in the mid 1990s. Instead of getting only the concession capacity of a decommissioned boat, the quota allocation of that boat could be partly retained for 13 or 18 years, depending on whether the decommissioned boat was destroyed or not.[17]

While in the Norwegian concession system fish quotas can be traded indirectly for the long term, the leasing of the current year's quota is not permitted. Even voluntary swapping is contrary to the regulations and punished if discovered. The economic loss of not allowing such practices could be substantial. One famous case involved shrimp quotas off the east and west coasts of Greenland. Two boats had shrimp quotas on both coasts, and their owners concluded that it would make most sense for one of them to go to the west coast while the other went to the east coast, instead of both going both places. This presumably involved substantial savings in time and fuel. Both knew, however, that this would be in contravention of the regulations and decided to report their catches as if they had been taken in both places. But they were found out and had to pay hefty fines. While the need to regulate fisheries is not in question one can certainly go too far in that direction; regulations that forbid saving time and money leave much to be desired.

Cod Fisheries

Entry into the fisheries for groundfish (cod and similar species) was mostly unregulated for much longer than in the pelagic fisheries, with the exception of the trawl fisheries, for which concessions have always been required. Total quotas were introduced in the cod fishery in 1977, after the 200-mile limit had been established and Norway and the Soviet Union had agreed on splitting the total permitted catch of Arcto-Norwegian cod in the Barents Sea and the Norwegian Sea evenly, with a small allocation to third countries that used to fish for cod in that area. The trawlers were put under a vessel quota regime, but trading in quotas was not allowed. No quota regulation was applied to the boats fishing with nets and lines, and the unregulated fishing by this part of the fleet often led to Norway exceeding its quota allocations. As the cod stock declined, the need to reign in the activities of these boats became more and more pressing. Initially this was done by stopping the fishery when the total permitted catch had been taken. This led to short fishing seasons and much competition for the largest possible share of the catch.

Finally, in 1990, when the quotas for the Arcto-Norwegian cod were at their lowest point ever, it was deemed necessary to allocate quotas to individual boats. The allocation was done on the basis of boat size; boats in a certain size class and which had caught more than a certain minimum of fish in the past all got the same quota and came to be referred to as Group I boats. Boats which had landed only small quantities in previous years got a common quota for which they could compete and came to be referred to as Group II boats. The distinction between groups with different quota rights was to linger on, as further discussed below. The quotas were not transferable.

An Attempt to Establish ITQs

On the basis of this experience it was attempted in 1991 to set up an ITQ system in the Norwegian fisheries. These attempts came to nothing, however. While the civil servants in the Ministry of Fisheries and some leading members of the Labor Party seemed reasonably convinced of the advantages of this proposal, it met with decisive opposition from other circles. The proposal was circulated among local governments and industry organizations and received a cool reception; negative reactions were more numerous than positive ones. The reactions provide an insight into the grounds for the opposition. Some took an explicitly ideological stand, opposing a privatization of a common resource and making it subject to market transactions for private gain. Some arguments were grounded in a fear that transferable quotas would generate migration of fish quotas from the northern to the southern part of the country. This "North-South" conflict is of a long standing. Fishing concessions for the purse seine fleet, which, as earlier stated, could be bought and sold, had over the years gravitated from the north to the south, and regulations had been put in place to make such transactions less profitable than transfers the other way around, albeit with little success. The idea that fish quotas would migrate from north to south was therefore not far fetched. Even if this would probably be advantageous from the point of view of economic efficiency, the spokespersons for the northern areas could certainly be expected to resist it.

A related argument was that people with easy access to money would more easily buy quotas than the less wealthy. This might encourage large-scale operations and absentee ownership. Sympathy for small-scale operations and opposition to absentee ownership has long been strong in the Federation of Norwegian Fishermen, an organization with a strong influence on policy. This has resulted in regulations limiting the ownership of fishing vessels to active fishermen. This arrangement is, however, impractical in the trawl fisheries, which have long been exempted from regulations limiting ownership to active fishermen. Although the sentiments against absentee ownership now seem on the wane, it still meets a strong resistance.

In sum, ideology and opposition to structural changes killed the proposal. Responses pointing out economic gains, for the industry or for the economy overall, were relatively rare. Fear of disadvantageous structural changes dominated expectations of favorable ones. The ideological element was strong in some political circles. There was strong grassroots opposition in the Labor Party as well as by some closer to the party leadership. It probably did not help that an application to join the European Community, as it was called at the time, had been submitted and was due to be voted on in a referendum a few years later. Opposition against membership was strong in fisheries circles, due to the Common Fisheries Policy. The Community does not recognize national 200-mile zones, and decisions on total catch from the various fish stocks and its allocation between nations are taken in Brussels, neither of which is popular among Norwegian fishermen. Antagonizing fishermen in the north would not have helped the referendum, which eventually came out against membership anyway.

Recent Developments: Toward ITQs?

As the Arcto-Norwegian cod stock recovered, the individual vessel quotas were abandoned for boats less than 28 meters long. Instead the total quota was divided between different classes of boats, partly according to size and partly according to the fishing record before 1991. The distinction between Group I and Group II vessels turned out to be there to stay; Group I vessels got quotas for different size classes of vessels, with an upper limit to how much each vessel could catch, in order to limit competition for the given quota. The Group II vessels got a much smaller common quota. Furthermore, entry into Group I was effectively closed while entry into Group II was still free. Not surprisingly, boats in Group I have turned out to have a value over and above what they would fetch for purposes of fishing only. Formally an acceptance by the regulatory authorities is required for transferring fishing rights when vessels change owners, but this has never been refused. The fishing rights of boats in Group I are valuable because such boats can catch more fish than those in Group II and the competition among them is less because of the closed entry. This management regime is applied to vessels less than 28 meters in all Norwegian fisheries, not just in the cod fishery in the north.

Meanwhile the pressure from boats less than 28 meters on the fish resources has increased. Technological progress has made these boats increasingly effective. As an example, small boats fishing for mackerel by hand line have now installed machines which jig the line automatically, and the fishermen can enjoy a cup of coffee until the time has come to pull in the fish. There has been relatively little control over investment in small boats, and not surprisingly their number has increased in response to increased profitability in the fishery. This has been true, in particular, of the mackerel fishery, the profitability of which rose substantially in the mid to late 1990s. The need to deal with this has become more pressing over the years. The industry has gradually come around to the idea that the total catch quota should be divided among different categories of boats. The pressure in this direction has come in particular from the owners of the larger boats which are under a concession regime and which get individual vessel quotas based on the vessel's cargo capacity or other vessel characteristics such as length. These industry players perceived a threat from the small boat fleet; the ongoing expansion of this fleet would in their view undoubtedly lead to demands for a greater share of the total catch quota, and such demands have in fact been heard from time to time.

In 1994 the Federation of Norwegian Fishermen agreed on a division of the total catch quota between what were called ocean going vessels and smaller coastal vessels. The limit between the two is usually set at 28 meters in length. The agreed division of resources between these two groups of vessels was controversial, and it was deemed necessary to deal with the problem more thoroughly and in a longer time perspective. In the late 1990s the Federation appointed a committee for this purpose. After difficult negotiations the commission agreed on a proposal, which in 2001 was accepted by a general assembly of the Federation. What is particularly noteworthy is that the commission emphasized the need to limit entry to the industry at all levels. Parallel to this, other developments were afoot. The time had apparently become ripe for a greater acceptance among fishermen at large not only for limited entry but for tradable fishing rights. In the late 1990s a system was put in place for vessels more than 28 meters long, by which a boatowner can buy the quota allocation of another boat and add onto his own, with certain restrictions and for a limited time. The explicit purpose was to promote restructuring of the fishing fleet toward greater profitability and less overcapacity.

In the summer of 2002 the Ministry of Fisheries came up with plans to introduce individual and transferable vessel quotas for smaller boats as well. According to this plan, there will be definite limits to transferability. Short-term transfers will be subject to authorization by the Ministry of

Fisheries, and long-term transfers of quotas will not be permitted for the smallest boats (less than 15 meters). The proposal has given rise to a heated debate involving not only fishermen but also politicians and academics. The opposition to these plans appears to come primarily from academic and political circles and to be grounded in ideology and a desire to avoid market driven processes and the structural changes these give rise to. In 2003 the Ministry's plans were endorsed by the Norwegian parliament.

The development in Norway has thus clearly been toward tradable fishing rights. Even if it has taken longer and been more hesitant than in many other countries, and there is a long way to relatively unfettered transferability like in Iceland or New Zealand, there is no mistaking about in which direction things are moving. To some extent the Ministry of Fisheries wishes to rely on market forces in its fishery regulations, and over the years considerable gains have been realized by so doing, particularly for the purse seine fleet. Nevertheless, a certain reluctance to rely on market forces and desire to retain ultimate control can be detected in the various restrictions put on transferability of quotas and concessions. The two metaphors which seemed most appropriate to two notable commentators describing the Ministry and its role were a fire department and a Soviet-type planning bureau. The silver lining of that cloud was that they came out in favor of the fire department.[18]

Canada

Before the 200-mile zone came into being, some foreign vessels took part in the sablefish and trawl fisheries off the coast of British Columbia, the Pacific province of Canada.[19] These vessels were displaced after the Canadian economic zone was established in 1977. The Canadians expanded quickly into the void (similar to what happened in Iceland), and limited entry was imposed in 1976 for the trawl fisheries and in 1981 for the sablefish fishery. Limited entry dealt only with the number of vessels but not with their use, specification and equipment. There remained both the incentive to expand effort and the ability to do so, through longer hours, larger boats, more gear, and better equipment. In the trawl fishery trip limits and monthly catch limits were imposed in order to keep the total catch within the overall annual limit, leading to at-sea discarding and misreporting. In the sablefish fishery the fishing season shrank from 245 days in 1981 to 14 days in 1989 despite the fact that the total catch quota increased by more than 40 percent and despite limited entry in terms of number of vessels. In 1995 the trawl fishery was closed for five months and reopened in 1996 with still tighter regulations.

Players in both fisheries took the initiative to negotiate with the government about ways out of the regulatory impasse, in order to salvage the industry from feared bankruptcy. A contributing factor was that the trawl fishery had to pay the cost for having observers onboard, which in fact pushed some of the smallest vessels out of business. The outcome of this was an individual vessel quota program, one for each of the two fisheries. These programs differ for the two fisheries and were negotiated at different times in the 1990s. The quotas in both programs are transferable but subject to different rules. The sablefish program is much simpler and its details were more easily agreed. The reason probably is that there were much fewer players in the sablefish fishery, or 48 against 142 in the trawl fishery. The public input also appears to have been much greater in the process leading to the quota program in the trawl fishery. That program grew out of a consultation process involving not only boatowners but also processors, crew, and community leaders, and is "exceedingly complicated," as one fisheries administrator puts it.[20] It is tempting to see this complexity as the result of a wide definition of stakeholders. Many of these have different and perhaps antagonistic interests and maybe idiosyncratic points of view as well. A common procedure in trying to reach agreement in situations like that is to satisfy different stakeholders by adding conditions that meet their concerns without totally alienating others. The result often is not just one of papering over differences but a program which, in terms of efficiency, is less than ideal and possibly dysfunctional. This is to an even greater extent a problem in the Alaska halibut quota program, to be discussed in chapter 7. A lesson one may draw from this is the instrumentality of defining stakeholder groups narrowly and in such a way that their interests are well aligned. Otherwise the result might be like the one emerging from the proverbial committee appointed to design a horse. It came up with a hybrid of a cow and a camel.

A part of the trawl fishery in British Columbia is directed at rockfish. The rockfish is a collection of different species and stocks, all of which are characterized by slow growth and survival to a high age if left unfished; the age of one specimen of rockfish has been estimated to 205 years.[21] These fish can be easily accessible when they aggregate to spawn. The risk of over-exploiting fish of this kind is high, and it is important that the exploitation rate be set low in order to avoid a possibly irreversible depletion.

Rice (2003) identifies several driving factors behind the ITQ regime for rockfish. One was the complexity of the regulations intended to extend the catching process over the entire year while still allowing competition for a limited overall catch quota. Another was a desire by parts of the industry to supply small but regular deliveries of high-priced fish rather than irregular spouts, which would glut the market and result in low prices and revenues. At the same time, there was opposition in the industry from those who felt they could outdo their colleagues in a competition for a given catch volume. Labor unions in the processing industry also opposed ITQs, because small quantities of high-priced fish are delivered to the fresh fish markets after a minimum of processing. This is reminiscent of the Alaska halibut fishery; waste often generates its own constituency among those who benefit from it and have an interest in perpetuating it.

Coincidental factors seem to have worked toward promoting the ITQ regime. In the 1990s there were deep crises in two major Canadian fisheries, the salmon fishery on the Canadian Pacific coast and in the cod fishery off Newfoundland. These crises led to increased government funding for restructuring in the fishing industry, which the Pacific rockfish fishery was able to tap into. This facilitated what originally was an individual vessel quota program, much like what happened in the fishery for Pacific halibut, with transferability coming later. The effects of the ITQ regime are substantial. In 2001 the value of rockfish landings by the trawler fleet in British Columbia was $37 million, versus $12 million in 1995, while the total catch had fallen from 21,000 tonnes to 17,000 tonnes. With the ITQ regime came greater involvement and contribution from the industry in the management of the fishery. In 1991 the industry paid a fee of $100 per license to the government; in 2002 the industry contributed $3.65 million to the management costs of the fishery.

While the gains from a greater efficiency in fishing were substantial for individual entrepreneurs in the industry, the rockfish fishery cannot be characterized as a major source of wealth at the provincial or even the local community level. The industry has never been a major provider of employment in the communities where it is based. This relative unimportance of the industry, except of course for the individuals engaged in it, has been advanced as a reason why the ITQ regime succeeded.[22] The ITQ regime could develop in relative obscurity from the public eye; civil servants and politicians did not have to worry about political repercussions from struggles over ITQs, and local community authorities were not overly worried about the loss of job opportunities as a result of a smaller fleet and less fish processing.[23] This conclusion runs counter to several strands of established wisdom; that the fishing industry must be sufficiently important for a country for politicians and civil servants to use time on improving its management; and that management reform should be as open and inclusive process as possible, with a comprehensive definition of stakeholders. In this case, however, the obscurity of the process and a narrow definition of stakeholders apparently helped to put in place a regime which is not only more productive but also contributes substantially to management costs and promotes sustainability of the resource by focusing the industry's interests on small but high-priced catches rather than large, low-priced volumes. In fact, two earlier attempts at introducing ITQs in the rockfish fishery failed because of an intervention by policy experts who felt that this would be a dangerous precedence for other fisheries in Canada. As already noted, the definition of stakeholders was narrower still and correspondingly more helpful in the sablefish fishery.

Perhaps the decisive factor in making the Pacific rockfish fisheries a success story is that these fisheries never had any role in the social policy of the province or the communities where they are based. This is a major contrast to Atlantic Canada, and in particular Newfoundland, where the fishery is seen as a way of life and a cultural phenomenon as much as or even more than a source of wealth. This misdirected benevolence ended in tears in the early 1990s when the Northern cod stock collapsed, not to recover since. This collapse is likely to have been precipitated by cutting the total catch quota too little and too late, in an attempt to maintain employment. Similar attitudes have long prevented a rational management regime from emerging in the fisheries of New England and Norway. In New England this has contributed to the demise of the fish stocks; in Norway it has manifested itself largely in excessive fishing capacity and employment in the industry.

The Pacific Halibut Fishery
The halibut fishery in British Columbia was put under an ITQ regime a little earlier than the sablefish and the trawl fisheries.[24] Here as well the

initiative came from industry circles together with key civil servants, although the fishermen were not unanimously behind it. The background was a formidable increase in the participation in the fishery and a resulting shortening of the fishing season, as the fishery was controlled by an overall limit on the catch. The number of boats participating in the halibut fishery increased from 333 to 435 from 1980 to 1989. Even if the catch almost doubled over the same period (it went from 5,560,000 to 10,470,000 pounds), the fishing season fell from 65 to 11 days, and further to just 6 days in 1990.

The ITQ proposal was opposed by the crewmembers' union and the fish processors. These groups probably anticipated that they had little or nothing to gain from the program, an expectation that turned out not to be unfounded. In 1990, just before its introduction, the ITQ program had the support of 70 percent of the fishermen and was initially put in place for a trial period of 2 years. At the end of the trial period the support had grown to 90 percent.[25] During the trial period, transferability of the individual quotas was not permitted. The effect of the individual vessel quotas was immediate and dramatic; after the program was put in place in 1991 the fishing season went from 6 days to 214 days. The value of the catch increased, as it could be absorbed by the fresh fish market rather than being frozen and turned into an inferior product.[26] Traditionally less than half of the catch was marketed fresh, but after the quota program was put in place that share increased to over 90 percent. The ITQ program made it easier to keep the total catch within the set limits. In the years 1980–1990 the total quota limit was exceeded every year except 1980. After 1990 the total catch has stayed well within the set limits. The industry itself pays for the monitoring and enforcement of the ITQ program.

The extension of the fishing season strengthened the bargaining position of the fishing fleet relative to the processors; captains could shop around for the best deal among the processors instead of having to land the fish in a hurry. This weakened in particular the position of the large processors; the number of processors in fact went up after the vessel quota program was put in place. The processors' opposition to the vessel quota program thus was not without foundation.

After the two-year trial period was over, transferability of quotas was allowed, albeit with certain restrictions, which have been relaxed somewhat later on. Transfers must still, however, be approved by the Canadian federal authorities. Only after transferability was permitted did the number of vessels participating in the fishery start to fall, from a peak of 435 in 1990, on the eve of the vessel quota program, to 286 in 1995. Demand for labor has also fallen, and so has the crew share of the catch. This may indicate a weaker bargaining position of ordinary crewmembers, but the incomes of fishermen have not necessarily fallen as well; sometimes at least the fall in crew share has been accompanied by fewer people employed on a vessel.

An econometric analysis concludes that the Canadian Pacific halibut ITQ program has resulted in significant economic gains, not just in terms of cost savings but also in enhanced revenues. It also concludes that constraints on transferability of quotas may be serious impediments to the realization of potential gains in efficiency.[27]

6   ITQs in Iceland: A Controversial Reform

As was earlier noted, in 1948, in the wake of the Truman Proclamations the Icelandic parliament passed a law that amounted to a claim to fish resources in the water column above the continental shelf.[1] Initially the Icelanders made no attempt to realize this claim. They waited for the verdict of the International Court of Justice in the case between Norway and Great Britain, to which the parties had agreed to submit their dispute. Norway claimed a 4-mile fishing limit and the closure of fjords and bays while Great Britain argued that the 3-mile limit was an internationally accepted rule. After the court had pronounced in Norway's favor Iceland extended her fishing limits to 4 miles and closed off fjords and bays (1952). Iceland, and Norway before 1815, had both been parts of the Danish state, which had claimed a wider jurisdiction at sea than 3 miles. In 1901 the Danish government had made a treaty with Great Britain, recognizing the 3-mile limit around Iceland.

The extension of the fishing limits brought Iceland into conflict with Britain. At that time Icelandic trawlers landed a large part of their catch in British ports, and British trawlermen imposed a ban on landings of Icelandic fish, with the support of the British government. This, however, had limited effect; the Icelanders were able to find new markets for their fish; fresh fish markets in Germany, and markets for frozen products in the United States and the Soviet Union.

Later Iceland was well ahead of the development in the international law of the sea, and may in fact have helped accelerate it. After the first UN Conference on the Law of the Sea in Geneva in 1958 had failed to recognize a fisheries jurisdiction of 12 miles, Iceland extended its fishing limits to 12 miles. This brought the Icelanders into conflict with the Germans and the Belgians, whose vessels fished off Iceland at the time, and with the English, who had been fishing there since the 1400s.

The United Kingdom, which still vigorously supported the 3-mile limit, would have none of this and sent frigates to prevent British trawlers from being apprehended by Iceland's coast guard. It was an uneven match: Iceland has no navy, only small and virtually unarmed coast guard vessels. The British were, however, civilized enough to restrain their use of force, and doing so would in any case have earned them international opprobrium. Iceland capitalized on the David-versus-Goliath metaphor, and fishing in organized groups under the protection of frigates was apparently not the best way to go about that business. After a few years, an agreement was reached, the frigates were withdrawn, the British trawlers were allowed to fish for a few years within the 12-mile limit, and Iceland undertook to submit any future disputes over fishing limits to the International Court of Justice. By that time, 12-mile limits were being increasingly accepted internationally, not just as fishing limits but as territorial limits at sea.

In retrospect the fuss about the 12-mile limit seems exaggerated. It is totally inadequate for establishing national property rights to most fish stocks, not just at Iceland but all over the world. Exceptions are sedentary and semi-sedentary species which live in shallow waters and do not migrate very far. Few such species were of any significance for Iceland at the time, and those that were (Norway lobster, for example) were not fished by the British. Cod, which the 12-mile limits were mainly about, migrate far outside those limits. But it made fishing a little more difficult for foreign trawlers, as the technology was at the time (the British trawlers were not very large and pulled in the trawl on the side, unlike the stern trawlers which the Germans were beginning to use). As a consequence, the Icelandic share of the cod catch went up, but the total catches of cod at Iceland have never been higher than in the mid 1950s, before the 12-mile limit was put in place. Figure 6.1 shows the catches of cod at Iceland since the early 1900s, both for Icelanders and foreigners, and the foreigners' share of the total catch. The two world wars are very visible, as is the effect of the 200-mile zone, but the 12-mile limit established in 1958 made only a small and temporary dent in foreign fishing and the total catch did not increase.

In the early and mid 1960s the Icelandic economy enjoyed a major boom from the herring fishery. This was due to a new fishing device (the power

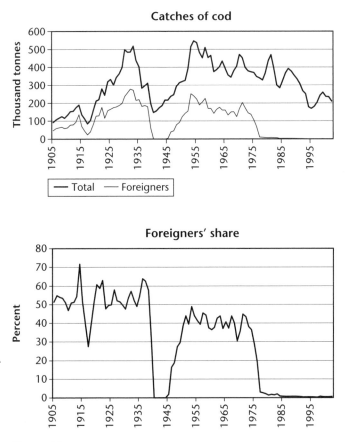

Figure 6.1
Catches of cod at Iceland and foreigners' share of the total, 1905–2002. Source: State of Marine Stocks in Icelandic Waters 2002/2003, Hafrannsóknastofnunin (Marine Research Institute), Reykjavík, Fjölrit no. 97, 2003, table 3.1.1.

block) and better fish-finding equipment, both of which were being introduced in this fishery. The power block made it possible to use much larger seines and boats, and the fishing capacity of the fleet was increased many times over. But the herring stock was decimated. The herring fishery was usually conducted outside the 12-mile limit, and boats from many nations participated in the fishery. No attempt was made at regulating the fishery, and in fact the fisheries biologists may have been too late in realizing what was happening to the stock. The herring typically occurs in large shoals, and the purse seine fishery is conducted by encircling a shoal and scooping up the fish. The ability to find and encircle a shoal is probably not much affected by the size of the herring stock, so the signal of falling catches per unit of fishing effort was not forthcoming until the stock had been nearly wiped out. Furthermore, the new fish-finding equipment made it much easier to locate the shoals. The boom ended in tears in the late 1960s; the herring disappeared, the Icelandic economy was afflicted by unemployment not seen since before World War II, and there was considerable emigration. To many people the most promising way forward appeared to be taking over the cod fishery where foreigners still used to take 30–40 percent of the catch.

In 1972, on the eve of the third UN Conference on the Law of the Sea, Iceland extended her fishing limits to 50 miles. For this there was no precedence in international law, the only rationale being that it coincided roughly with the 200-meter depth contour usually taken as the limit of the continental shelf. Britain protested and took the case to the International Court of Justice, as provided for in the agreement reached between Britain and Iceland in 1961. Iceland did not bother to participate in the court's proceedings. Britain sent in her frigates again but this time the Icelanders had changed tactics, cutting the trawl wires of the British. In 1974, after many dramatic incidents at sea, an interim agreement was concluded. That, however, did not hold very long; at the Caracas session of the UN Conference on the Law of the Sea in 1974 it became clear that a 200-mile economic zone had wide support, Iceland extended her fishing limits to 200 miles in 1975, and a third cod war broke out. It was eventually defused a year later through intermediation by the Norwegian government. The British were allowed to fish a small quantity for a few years, but by 1980 foreign fishing for cod at Iceland had virtually ceased.

Toward ITQs

With the foreign fleets having been driven away, the Icelanders could manage their cod fishery, and other groundfish fisheries within the 200-mile limit, as they best saw fit. Even if they had argued their case for extended fishing limits by the need to limit the pressure on the cod stock they were slow to apply that logic to their own fishermen, and even slower in managing their newly acquired resources in an economically sensible manner. Instead they enlarged their fishing fleet, bought new and well equipped trawlers, and expanded into the void the foreigners had left behind. They did, however, put in place a limit on the total catch, aiming at rebuilding the cod stock to increase long-term catches. The stock recovery failed to materialize, however, and the fishing fleet soon turned out to be too big for the total catch quota being set. Attempts were made to keep the catch within the quota limit by allowing the fleet to fish for cod only a limited number of days per month. Even so, in some years the quota was exceeded. The quota was probably too high to achieve any stock rebuilding; it was usually set higher than recommended by the fisheries biologists assessing the stocks and advising on how large the total quota should be.

By 1983 the cod stock had reached an all-time low. There was a sense of crisis, in the industry and the country at large. The Minister of Fisheries proposed a system of individual vessel quotas for cod and some other types of groundfish, to be allocated on the basis of recent catch history. The proposal was debated and finally accepted by the "fisheries parliament," an annual meeting of elected representatives from the industry, both the fishing and the processing side. Reports on the debate at the time, both in the "fisheries parliament" and elsewhere, make it clear that the quota allocation scheme was seen by many, and probably the majority, as a temporary measure to deal with a hopefully temporary emergency, to help rebuilding the depleted cod stock; in fact some knowledgeable people maintain that the individual quotas would never have been accepted by the industry except as a temporary measure dealing with an exceptional emergency.

The fact that individual quotas were proposed as a means to rebuild the stock is slightly surprising, as a low enough overall catch quota would be enough to deal with that problem. But restricting the number of fishing days had proven inadequate to limit the catch as required, and vessel quotas were seen as a more effective measure. The quotas were made transferable, in order to increase flexibility and efficiency, since there was too little fish to allow all boats to be used at full capacity. There does not appear to have been much debate at the time about the transferability of quotas. Needless to say, transferability of quotas valid for only one year does not provide much of an incentive to reduce fishing capacity. Few boatowners would trade in their boats for smaller ones if their quota is temporarily small and they expect to have the opportunity to catch more in the not too distant future, and the market value of quotas valid for just one year would hardly offer sufficient compensation to lay up a boat permanently and withdraw from the industry.

Individual transferable quotas were in fact not entirely novel in Iceland when they were introduced for the first time in the cod fishery. Some years before they had been put in place in the herring fishery and the capelin fishery. The herring fishery collapsed, as earlier mentioned, in the late 1960s and was banned for a number of years after that. A small, local herring stock at Iceland had survived the debacle and was in a sufficiently good shape for a small catch to be permitted in 1975. The catch was much too small to be meaningfully parceled out among the boats which traditionally had participated in the herring fishery, and a way had to be found to limit the number of boats. Some administrator came up with the idea that requiring the herring to be salted onboard would do; many boats were too small for having that kind of operation done on board. The eligibility rules were framed accordingly, but some bright boatowner realized that the rules did not prevent the herring from being salted on board when the boat was back on the wharf. Therefore, the rule was not quite as effective as expected. This ridicule, and others, paved the way for more rational rules such as dividing the herring quota among the boats and making the boat quotas transferable so that everyone eligible could get a share of the pie while the fish was being caught in an efficient way. The capelin fishery took off after the herring fishery collapsed and became the main source of meal and oil, as the voluminous herring catches had been before. It is noteworthy that the capelin quotas were not allocated on the basis of catch history but according to a formula giving two-thirds as equal shares to the 52 eligible vessels, dividing up the remainder according to the individual vessel's cargo capacity.

The quota system introduced as an emergency measure in the cod fishery in the fall of 1983 (effective from the beginning of 1984) came to stay. The rebuilding of the cod stock was a slow process, and after the first year with individual boat quotas the capacity of the fishing fleet was still much too high. The quota system was prolonged for 1985, then was prolonged 2 years in 1986, and then was prolonged 3 years in 1988. Along the way several and sometimes counterproductive modifications were made. The quota management system met an increasingly vigorous resistance from some boatowners, especially from a part of the country known as the Westfjords. The fishermen in this area felt they had a competitive edge in being close to some of the richest fishing banks in the country and that the quota system barred them from taking advantage of this. Fishermen with recently acquired boats felt disadvantaged because of their short catch history. To accommodate these views a dual system was operated in the period 1986–1990; the ITQs and a system known as effort quotas where a boat was allowed to fish for a certain number of days to improve upon its catch history, but with an upper limit on the cod catch that could be taken. The effort quotas, which were particularly attractive in 1986 and 1987, had undesirable effects on efficiency. Investment in boats increased, in a scramble to get a larger share of the pie. That could only happen, however, at the expense of those who were "loyal" to the quota system.

Another mistake being made when the quota system was initially put in place was to exempt boats less than 10 GRT (gross register tons). In 1983 there were only a few of these boats, taking about 3 percent of the total catch of cod, and it was felt that leaving them out would not much matter. The consequences were dramatic and, it would seem, predictable. Over a few years the number of small boats multiplied many times over, and by the mid 1990s their share of the total cod catch had come to exceed 20 percent. In the meantime various ways had been tried to limit their catch, mainly by restricting the number of fishing days and reducing the maximum size of boats in this category to 6 GRT, but with limited success; entry into this sector was still open. Owners of small boats formed an articulate pressure group that could count on substantial sympathy from politicians and the public alike (David versus Goliath again). Owners of small boats would occasionally stage demonstrations outside the parliament building or flock to the public gallery where its proceedings could be followed.

Finally, sufficient support emerged, both in political and industry circles, for the quota system to be put on a firmer footing. In 1990 a new fisheries law was passed, removing the time limit on the quotas and making them valid for the indefinite future. They are not, however, explicitly permanent as the case is in New Zealand. The small boats were supposed to be dealt with within a year but the Icelandic politicians pushed that problem before them for about 10 years; finally in 2001 most of the small boats were incorporated into a quota system of their own. Owners of small boat quotas cannot sell them to larger vessels while transactions the other way around are permissible.

Achievements

What has the Icelandic ITQ system achieved and what could it be expected to achieve? Some commentators have noted that it has not resulted in a recovery of the cod stock, the single, most important resource for the Icelandic fishing fleet. This can be seen directly from the stock estimates published by the Icelandic Marine Research Institute (figure 6.2) and also gleaned from figure 6.1 which shows that the catches of the Icelandic cod stock peaked in the mid 1950s and have trended downward ever since. This is a misplaced criticism; resource conservation is achieved by setting the total quota appropriately. ITQs are mainly a tool to achieve economic efficiency, even if they should help conservation by making it easier to keep the catch within set limits and by fostering a conservation attitude in the industry. It appears that the total catch quota has simply been set too high. Since this is a decision taken by the Ministry of Fisheries, the responsibility lies squarely with the government and the political parties behind it; no lesser authority than the Prime Minister himself has used prospects for a greater catch quota based on an expected stock recovery as an argument in election campaigns. One could have hoped for more effective pressure from the industry for setting catches more conservatively in its own long-term interest, but either that pressure has not been forthcoming or not been effective enough. That the catches have been too large

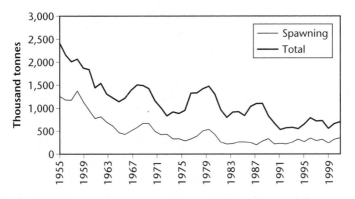

Figure 6.2
Estimated size of the Icelandic cod stock, 1955–2002. Source: State of Marine Stocks in Icelandic Waters 2002/2003, Hafrannsóknastofnunin (Marine Research Institute), Reykjavík, Fjölrit no. 97, 2003, table 3.1.10.

can be seen from the mortality they have produced in the stock; in the years 1984–2002 the fishing mortality for cod 5–10 years old was no less than 0.73 on the average.[2] With an assumed natural mortality rate of 0.2, this means that a year class of fish will at the end of each year have been reduced to 40 percent of its size at the beginning of the year.[3] Other possible reasons for the stock decline are a change in ocean climate and the development of the capelin and shrimp fisheries, but both of these are important food sources for cod.

In the view of this author, there is little doubt that the ITQ system has increased the efficiency in the Icelandic fisheries, even if there may still be some efficiency gains to be realized. Figure 6.3 shows the total amount of fixed capital in the fishing industry (the catching sector) in Iceland each year since 1970. A crude impression of productivity can be obtained from comparing this with the total amount of fish caught. The total catches of fish and the total amount of fixed capital grew roughly in parallel from 1970 to the mid 1980s. There was a steep growth in capital invested in the fishing fleet in the early 1970s, in the wake of the expansion of the fishing limits to 50 miles. Another splurge in investment in fishing boats occurred 1986–1988. It is, to say the least, tempting to see this latter increase as a

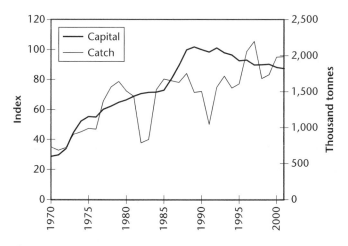

Figure 6.3
Total fish catches of the Icelandic fishing fleet and total capital (index) in fish catching, 1970–2001. Sources: Útvegur, Statistics Iceland; National Income Accounts, National Economic Institute of Iceland.

consequence of the effort quota system discussed above, but some of it may have been due to the increasing use of trawlers that process the catch on board, which are more capital intensive than trawlers which supply fish to shore-based processing plants. Since the late 1980s the amount of capital invested in the fleet has stagnated or even declined, despite a continuing increase in the total catch. It is probably no coincidence that this happened when the ITQs were put on a firmer footing and made valid for the indefinite future. Only ITQs that are valid for the long term can be expected to remove the incentive to invest in boats and equipment solely for the purpose of getting a larger share of a given total. The fall in the total amount of capital has not been accompanied by an increase in the use of labor; the total amount of labor used in fish catching in Iceland has been fairly constant for the last decade or more.[4]

Others have been less sanguine in their conclusions. A review of the quota system in the late 1990s found that the profitability of the fishing industry had improved since the early 1980s, immediately before the quota management system was initiated (but, as earlier stated, it was fairly experimental until 1990).[5] The authors were reluctant, however, to ascribe this to the quota management system, at any rate exclusively, pointing out that the early 1980s were exceptional for the fishing industry and that after this period the profitability of this industry followed a pattern similar to other industries.[6] But this is not the whole story. Improved efficiency in an ITQ-managed fishery will be reflected in a rising value of quotas, both rental fees and as permanent assets. The asset value of quotas will be a part of the firm's capital and thus appear in the denominator of the expression for return on capital; in the long term the returns in the fishing industry would be expected to be the same as in other industries, irrespective of whether the fisheries are managed by quota or not managed at all. As figure 6.4 shows, the price of cod quotas in Iceland has increased over the years, indicating an improvement in profitability.

The quota prices shown in figure 6.4 most likely exaggerate the profitability of the industry, even if they certainly indicate that it is positive and improving. As figure 6.4 also shows, the quota prices amount to a very high share of the raw fish price of cod, on some measures even more than 100 percent. There are two possible and not mutually exclusive explanations for this strange phenomenon. First, those who rent quotas are the ones who supply fish to the most profitable markets. The most profitable

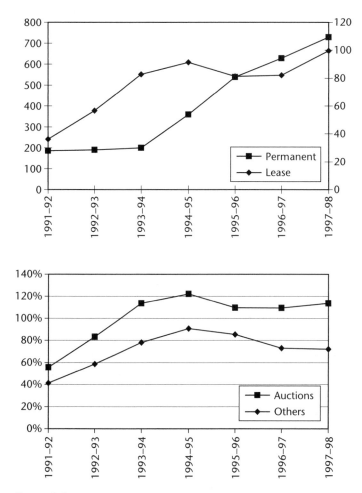

Figure 6.4

Upper panel: Lease prices of cod quotas in Iceland, in krónur per kilogram of gutted weight. Lower panel: Lease prices in percent of raw fish prices on auctions and in other markets. The numbers are averages for each "quota year," which begins September 1. Courtesy of Ásgeir Daníelsson, Iceland National Economic Institute.

markets are the fresh fish markets, while fish being processed into frozen or salted products cannot sustain equally high raw fish prices. Second, the leased quotas could be mainly additions to a larger "core" quota. A person who leases quota in addition to his own would not take into account fixed costs; these costs would have to be paid anyway, and the additional quota would be profitable as long as its lease price is less than the value of the additional catch it provides less the additional avoidable cost of fishing it. Hence the amount paid for the quota could easily be in excess of the rent it produces. It is noteworthy that the price of permanent quotas has sometimes been as low as 3.5 times and always less than 10 times the rental price. This indicates that the current rental price is higher than the expected future annual rent; if the current rental price were a correct reflection of the future annual rent the price of the permanent quota would be equal to the annual rental price divided by the real rate of interest, and we would not expect the latter to exceed 10 percent.

Yet another way in which the quota system is likely to have benefited the industry, and the Icelandic economy at large, is through the emergence of large, vertically integrated fishing companies. This has happened through certain companies buying up others, including of course their quota allocations. Some see this as an undesirable concentration, but in terms of economic efficiency it is likely to be beneficial. With less than 300,000 people living in the country the Icelandic economy is extremely small and so are its fishing companies; they do not come anywhere near the Fortune 500. The Icelandic fishing companies sell their products into foreign markets in fierce competition not just with other sources of fish supply but also in competition with other foodstuffs. It would be surprising if there were not some economies of scale and scope in an industry like that.

Controversies

Since 1990 and, for the small boats, 2001, the Icelandic ITQ system appears to be on a fairly permanent and secure footing. But there is still substantial, and perhaps even growing, controversy surrounding the system which puts its future in some doubt. The main controversy is about who owns and, by implication, who should benefit from the fish resources around Iceland. The fishery law of 1990 states that the fish stocks within the 200-

mile zone are the property of the Icelandic nation. Some legal experts have argued that this phrase is largely meaningless. They maintain that a statement like "the fish stocks around Iceland are the property of the Icelandic state" (presumably held in trust of the Icelandic people) would be quite different, since the state is an organized entity which can enter into contracts, enforce the law, etc. This, incidentally, is the formulation in the Norwegian petroleum law, which states that petroleum underneath the seabed of the Norwegian continental shelf is the property of the Norwegian state.

Nothing, so far as this author can see (and he is not a legal expert), prevents the state from putting in place private, transferable and permanent use rights, even if the fish stocks are classified as national or state property. The legislation in New Zealand has from the beginning been explicit on this point, which however did not prevent the government of New Zealand from radically altering the quota system a few years after it had been put in place, as discussed earlier. In the United States it is explicitly stated in the regulations establishing individual quota systems that they do not amount to property rights in perpetuity, presumably in order to exonerate the government from having to pay compensation for lost property in case it chooses to change the quota system. For practical purposes the difference between explicitly recognizing individual quotas as property and not doing so may be more apparent than real. In the United States and Iceland individual quotas are traded for the long term, apparently in the expectation that they will be valid for the long term (note that even with a moderate rate of discount, the present value of a quota 20 years hence is only a fraction of its expected nominal value at that time). Presumably, quota holders in Iceland and the United States could sue their government for damages if it changes the quota management system abruptly and dramatically.

The "property of the nation" formulation is, not surprisingly, frequently invoked by those who would like to see the government capture some of the rent that has emerged in the Icelandic fisheries. The nation would, as a landlord (or sealord), be entitled to a fee for allowing somebody to use her property. This view has substantial following, stimulated by the high quota prices.

The rent tax idea has been vigorously resisted by the boatowners who hold quotas and equally forcefully pursued by some economists and members of the general public. In 1999 a committee was appointed by the Icelandic parliament to investigate the issue of taxing the use of natural resources. Fish are not the only natural resource of interest in Iceland; hydroelectric power and geothermal energy (as water to heat houses and as steam to generate electric power) also are important and valuable natural resources, but fish was very much in focus. The committee did not reach consensus but pointed out two possible ways to tax fishing rents, a fee in some form or annual auctions of a certain fraction of the quotas. Yet another committee with a mandate to revise the law on fisheries advocated a fee rather than auctions. A law was passed in April 2002 imposing a fee on holdings of fish quotas. The fee is set to escalate over a number of years but does not seem likely to amount to much more than covering the public expenditures on fisheries management in Iceland, leaving little scope for rent capture. The fee is not levied on the profit of individual firms but calculated on the basis of assessed, industry-wide profits (revenues less labor and fuel costs) and levied on individual quota holdings, which avoids the pitfalls of taxes on the reported profits of individual firms.

Another controversy has concerned access rights to the fisheries. Not everyone appreciates closing access to a public resource which is, as stated in the fisheries law, the property of the nation. All would probably recognize the need to protect the fish stocks; without fish there will be no fishery. There is, however, considerable disagreement about how to do so. The experts at the Institute of Marine Research have argued for years that the catch of cod needs to be limited in order to allow the stock to grow. In recent years they have been more successful than earlier, as their recommendations on the total allowable catch have largely been heeded. The stock has, however, not recovered as expected (see figure 6.2), and the catches from the Icelandic cod stock have been on a declining long-term trend since the mid 1950s (figure 6.1). This has put the Institute on the defensive against those who argue that the catch quota has been set too low rather than the opposite, and that stock recovery has been impeded by a limited food supply causing large cod to prey upon smaller specimens of their own kind. This is not just a layman's opinion; some biologists have supported this view.

But when it comes to limiting the access to the industry, be it through individual quotas or in some other way, the consensus ends. Not everyone is equally appreciative of why this needs to be done. The argument can be paraphrased as follows: the fish are a common resource; it is even written into the opening paragraph of the fisheries law that the fish resources are the property of the Icelandic nation. The fish are the basis of our nation's existence, and it is outrageous to deny young people the opportunity to use these resources to support themselves and their families.

What this argument neglects is that no new wealth is created by having more people than needed going after the fish. It is the productivity of nature that decides how much fish we can take without putting the future of the fishery in jeopardy. Maximizing the economic welfare of the nation requires that her manpower and capital not be wasted in the fishery but directed to a better use. The problem in an unregulated fishery is, however, that the marginal boatowner would have an incentive to participate in the fishery even if he contributes nothing to the sustainable value produced in the fishery or makes it less than it could be. An additional fishing boat will catch some fish that other boats could have taken and may even reduce the surplus growth of the fish stock. If we subtract these negative effects affecting other boats from the catch value of an additional boat, we end up with less, and perhaps a lot less, than the value of the catch the additional boat will take by itself.[7]

Opening up the access to the Icelandic fisheries would thus achieve nothing other than retard the growth of the Icelandic economy and make it more likely that it will lag behind the neighboring countries with which the Icelanders are now roughly on par in terms of standard of living. It would blunt the incentives people now have to develop other sectors of the economy, which is the only way living standards in Iceland can be maintained and hopefully increased for a population that is still growing.

Legal Challenges to the Quota System

The debates on the merits, and demerits, of the Icelandic quota system have reached all the way to the Supreme Court of Iceland. Two cases stand out. The first case, resolved in early December 1998, resulted in a confused and contradictory verdict that apparently papered over disagreements among the members of the court. Like other texts marred by inconsistencies and contradictions this one inspired lively interpretations which went in various directions. The latter case, resolved in early April 2000, resulted in a three-way split among the seven judges of the court and revealed the deep disagreements among them.

The first case was initiated by one of the most outspoken critiques of the quota system. He had applied for a permit to fish and an allocation of specific amounts of various types of fish managed by quota. The Ministry of Fisheries dismissed the application on the grounds that the applicant did not own a fishing boat. According to the fisheries law in force at the time, permits for commercial fishing could only be given to boats which held such permits from before and had not been decommissioned, and to small boats (less than six GRT) built before a certain date. The plaintiff argued that limiting fishing permits to those who happened to own boats at a certain point in the past violated his constitutional right to pursue any occupation he desired.

A lower court initially threw out the case on a formality, but it was referred to the Supreme Court, which ordered the lower court to pass judgment on the substance of the case. The lower court then ruled that the Ministry's refusal was not in violation of the Constitution. The plaintiff referred the case to the Supreme Court, which ruled in his favor.

The Supreme Court's ruling came as a surprise to many people, not least because it was contrary to the ruling in the lower court. For a while the court's verdict greatly bolstered all those who opposed the quota management system on the basis of its perceived injustice. Some sent gleeful e-mails to their friends and acquaintances at home and abroad trumpeting that the quota management system was unconstitutional, foreseeing its unraveling in the not too distant future. That reaction was indeed understandable. The court's verdict contains much rhetoric to the effect that the quota management system as such is unconstitutional. The central conclusions of the court can be concisely summarized as follows[8]:

Since 1983 fishing permits have been limited to boats that were used at a certain time in the past. The owners of such vessels have from the beginning had a stronger right than those without command of such vessels. The plaintiff has accepted that initially, when it was judged necessary to limit the catches, it was not unreasonable to allow only boats with a certain catch record to fish. Such temporary measures should not, however, be made permanent, as in fact has happened. According to Paragraph Five in the fisheries law, the right to fish is tied to the ownership of vessels which were in use in the early 1980s or replacements of such vessels. As a consequence, no one has a right to fish commercially except those who at a certain time owned boats or later acquired such ownership through inheritance, purchase or in some other way. In 1983 the parliament concluded that specific measures were needed because of dwindling fish stocks at Iceland. At that time the method of allocating catches among boats was introduced, and such has been the practice ever since. It is unavoidable to conclude that this method implies discrimination between those who derive their fishing rights from ownership of fishing vessels at a certain point in time and those who are not or cannot be in that position. Even if temporary measures of this kind could have been justified to prevent a collapse of fish stocks, an issue on which the verdict does not pronounce, it is difficult to see any logic that compels legalizing for the indefinite future the discrimination which follows from Paragraph Five. The defendant (the Ministry of Fisheries) has not shown that other methods are not available to protect the fish stocks around Iceland. This paragraph is a significant hindrance for a large part of the nation to enjoy the same rights to the common property fish stocks around Iceland as do the relatively few persons who owned fishing boats when the quota management system was introduced.

From an economic point of view, these arguments are surprising and disturbing. The need to limit fishing is apparently seen as temporary, associated with the risk of stock crashes, presumably an infrequent emergency. Nothing is said of the need to limit access to fish as scarce resources in order to ensure maximum economic benefit for the nation. The fish stocks are referred to as common property, and even if not explicit the language certainly encourages the notion that common property implies that the object (the fish stocks) be accessible to the widest possible circle of people. The court made a major point of the fact that the right to fish was contingent on owning a fishing boat. Fishing boats can, however, be bought and sold, and there was a lively trade in them in Iceland in the 15 years between the establishment of the quota management system and the court's verdict. If the access to fishing is limited the access right becomes valuable and is likely to be traded for money or other favors, indirectly if not directly; not everyone can possibly get a right of access if such rights are to be meaningfully restricted. Allocating such rights through some form of market mechanism would be the prescription recommended by economists. Tying rights to boats and making the trade in rights partly contingent on trade in boats is perhaps not the best way to organize such trading, but there is no principal difference between buying a fishing right directly for a sum of money and having to buy a fishing boat in order to be eligible to buy such rights. In both cases the access is restricted to those who are willing and able to put up the sum of money which the owner of the right is prepared to accept.

Not surprisingly, the verdict gave rise to a lively debate and interpretation. Some concluded that it simply branded the quota management system as unconstitutional. From the language used in the verdict it is difficult not to sympathize with that interpretation. Others had a much narrower and more technical interpretation.[9] These commentators, most or all of whom were lawyers, pointed out that the court had only referred to Paragraph Five of the law and not to Paragraph Seven. Paragraph Five stipulated that permits to fish could only be given to boats while Paragraph Seven said that access to all fish stocks not managed by an overall catch quota was open for all with a fishing permit, but that the total permitted catch from stocks managed by a limit on the total catch should be allocated to individual boats.[10] Therefore, the court had said nothing about the constitutionality of the quota system as such, only that tying fishing permits to boats was unconstitutional.

This interpretation is technical and far fetched. Without the quota allocation stipulated in the said Paragraph Seven the right to fish would be of limited value, albeit not entirely valueless, since there were some stocks to which access was still free. The plaintiff had also asked for specific allocations of fish under quota management, an issue which the court did not address. The court's rhetoric applies no less and probably more to fish stocks under quota management than those outside; the latter are left out because the interest in catching this fish is so moderate that it does not seriously conflict with the constraints of nature. The alleged distinction between the two paragraphs was not exactly made sharper by the fact that the court used the word for fishing permit found in Paragraph Seven while allegedly dealing with Paragraph Five, which used a different word.[11]

Irrespective of what might be the court's meaning, the government had to do something about the constitutionality of the fisheries law, in particular Paragraph Five. The immediate consequences of the verdict were more than a little ironic, given its rhetoric. Anyone who so desired could now get a fishing permit without owning a boat. That did not mean an automatic quota allocation—Paragraph Seven had not explicitly been judged unconstitutional—but the small boats (less than 6 GRT) were exempt from the quota management system; they did not need individual quotas and could compete with other similar boats for a total allocation assigned to these boats (and which was typically exceeded). In order to prevent an undue increase in the number of small boats the fisheries law was amended in January 1999 to incorporate these boats more fully into the quota management system and otherwise limit their fishing activities. The most immediate effect of the verdict thus was to further restrict entry to the fisheries, in an apparent contradiction of the rhetoric of the verdict lamenting the barriers erected for a large part of the nation against utilizing her common property resources. Nothing was done about the quota management system as such, which the legal experts of the government considered not having been called into question by the court.

It did not take long until the Supreme Court got an opportunity to further clarify its thoughts about the quota management system and Paragraph Seven. A couple of months after the said verdict a fishing boat left harbor without any quota allocation to cover the catches its skipper intended to take. The skipper and the boatowner argued that they had intended to lease the necessary quota allocations while the boat was under way to the fishing grounds, but that this had turned out to be more expensive than expected. They argued that the quota allocation system was unconstitutional, in that the total catch quota was decided by regulation and not by law as required when restricting people in their choice of occupation. They argued furthermore that the discrimination between those who had and those who did not have a quota allocation was unconstitutional. Finally, they argued that the Supreme Court had implicitly found Paragraph Seven of the fisheries law unconstitutional.

This time the court split in three parts, with the majority of four seeing no conflict with the constitution, one minority of two judging the quota management system unconstitutional, and a minority of one in between but leaning toward the other minority view. In their verdict, the majority clearly emphasized that fishing needed to be limited, in the general interest of the nation. Limitations on fishing were not seen as exceptional measures necessary only for dealing with emergencies of impending stock collapses. The majority found it reasonable to allocate catch quotas to individual boats on the basis of their catch record. They also found transferability of quotas, in the long and the short term, in accordance with the first paragraph of the fisheries law defining the fish resources as the property of the nation, in that it would stimulate an effective use of these resources. The majority emphasized that the fish quota allocations did not amount to property rights in perpetuity but were of indefinite validity, as the quota system could only be changed by law.

The minority of two recognized the need to limit access to fish, apparently not just as a temporary measure to deal with emergencies but as a way to ensure rational utilization of the fish stocks. The minority considered Paragraph Seven of the fisheries law unconstitutional in that it restricted access to those who had owned boats in the early 1980s or acquired such boats or their replacements at a later date. This minority found it unacceptable that others had to lease access from these privileged few and argued that over time this was bound to lead to the emergence of privileged groups. The minority was silent, however, on how otherwise to resolve the dilemma of having to allow access for some and deny it to others.

The minority of one did not find the fisheries law unconstitutional. This judge also agreed that the quota management system was not in conflict with the first paragraph in the fisheries law, in that it promoted efficient use of the resources, but argued at length against the privileges bestowed upon a selected few by limiting quota allocations to boatowners. While this might be understandable as an emergency measure for a year or two, as the fisheries laws before 1990 had been, it was not defensible as a long-term solution and probably not sustainable either. The judge found it particularly unacceptable that newcomers to the industry had to buy their entry from those who had been active in the industry at a certain point in time.

The verdicts of the Supreme Court probably are a good reflection of the main challenges to the Icelandic quota system. The main concern is the perceived injustice in the distribution of the benefits of the system. People who initially got their fishing rights for free have been able to enrich themselves by selling or renting out those rights to others. This is becoming increasingly difficult to redress; the initial allocation of rights took place over 20 years ago, even if it has been tinkered with after that, so that many of those who now are active in the industry have bought their way in. It is doubtful, however, whether the process could have evolved differently; it is at least easy to understand why things happened the way they did. In 1983, when the quota system was first put in place as a temporary measure, the profitability of the industry was low and the quotas were not very valuable. The quota system was mainly supported by the industry itself and it was a close call; there is little doubt that quota auctions or quota fees would have been unacceptable to the industry at that time. In fact, in 1983 and 1984 the industry made substantial losses. Later opportunities to limit the windfall gains by the original quota holders were missed even if the rise in quota values was by then a well established fact. Here much blame must be laid at the door of parliamentarians who did not deal with the issue but used their time and influence to secure favors for owners of small boats, either for political gain or because of misdirected benevolence. There is no doubt that the leadership in shaping the quota system has come from the Ministry of Fisheries and the industry itself. The legislative assembly has followed up with the necessary legislation, but often with amendments which have been harmful rather than helpful. This lack of leadership from the legislature can probably be explained by the fact that most if not all political parties have been divided on the issue.

The quota management system itself is less controversial than its effects on the distribution of income. There are, however, many who do not see the need to limit access to the fish stocks, and even those who would doubt the need to limit fishing at all, pointing to the failure to rebuild the fish stocks despite more than 20 years of a quota regime. Then there are those who think that other methods than quotas would be better to limit fishing, such as limitation on the number of fishing days and direct limits on the number and size of boats, or perhaps competition for a given total catch. These methods are well tried in various parts of the world and have produced undesirable economic consequences.

The most controversial effects of the quota management system, apart from its effects on the income distribution, are its alleged effects on the structure of the industry. There has been considerable consolidation of quota holdings, and some fishing firms have grown quite large, partly by buying up or merging with other firms. This has had the effect of closing down fishing plants in certain communities and moving the operational base of some boats elsewhere. As a rule, small communities have suffered more than larger ones. This is easily blamed on the quota system; when a quota is sold permanently out of a certain community the local plant is left without its traditional fish supplies and people are thrown out of work. This is, however, probably an unavoidable and indeed desirable consequence of economic development. Fish processing is a typical low wage industry and not particularly attractive in other ways; most people who have a choice shun it, and it is not uncommon in small and isolated villages in Iceland to rely on labor from other countries with temporary work permits. Fish processing is increasingly taking place at sea, with a correspondingly declining need for land-based processing. Small and isolated fishing communities are not particularly attractive, especially for young people, because of few and uniform work opportunities and poor availability of services. The fact that quota holders do not elect to base their boats and fish processing in small, fish-dependent communities is an unmistakable sign that these communities do not possess any comparative advantage for this activity. The quota system in Iceland may very well have accelerated the decline of these communities, but it would probably have happened anyway, unless the Icelandic economy had stagnated or backtracked. Seen in that light, the structural changes following the quota system are to be welcomed rather than deplored.

7 The Development of ITQs in the United States

The movement toward rights-based fisheries management in the United States has been slower and more patchy than one would expect in a country that prides itself of an unrivaled economic strength based on free markets and private property. In fact, a misconception about free markets may occasionally have hindered the emergence of such rights. Some participants in the US debate have from time to time seen access controls to fisheries as interfering with freedom of enterprise. Freedom of enterprise should not be confused with free and uncontrolled access to a limited resource. Freedom to establish a lumbering business does not mean freedom to cut down trees at will without having bought the land or being otherwise authorized to do so. Freedom of access to a limited resource does not augment national wealth and usually not individual wealth either. "Olympic fishing," or "derby fishing," as open competition for a given fish quota is sometimes called, may be an appropriate way of organization if we view fishing as a sports contest but not if we view it as a wealth generating activity. Athletic competitions are not about getting to a specified goal with a minimum of effort.

The management of the fisheries in the federal waters (i.e., outside the old 3-mile limit) of the United States was designed by the Fishery Conservation and Management Act of 1976 as a transparent and democratically accountable process.[1] It is overseen by regional Fisheries Management Councils, of which there are eight, with members nominated by the governors of the states involved and appointed by the Secretary of Commerce. These councils hold open meetings and hearings and are charged with formulating a management plan for the fisheries in their respective region. The final approval rests, however, with the Secretary of Commerce, and it has happened that plans approved by the councils have not been approved by the Secretary.

It is possible that the emergence of exclusive use rights in the fisheries of the United States has been hindered rather than helped by the council process. Ordinarily, innovations in fisheries management such as individual transferable quotas have been introduced through the councils, but their passage has typically been a long, drawn-out, arduous process, which often has failed to reach fruition. Different interest groups have been preoccupied with trying to improve their position or to guard against a possible loss from a new management regime. The first and highly successful ITQ regime established in the surf clam and ocean quahog fisheries led to substantial structural changes which benefited some more than others, and yet others not at all. Later debates over ITQ regimes or other novelties in management have increasingly focused on preserving the relative position of different regions, fleet groups, or capital owners versus crew. The IFQ (individual fish quota) program in the Alaska halibut and sablefish fisheries shows this very clearly, and the plan for fish and fish processing quotas in the Alaska crab fisheries currently (2003) under consideration even more so. In recent years the US Congress has begun to act over the councils' heads, sometimes at least through initiatives from groups within the industry, which have found this an easier avenue than the council process.

The fisheries management councils can be viewed as mini-parliaments for fisheries management. The open meetings and hearings on various issues are in many ways a model of transparent decision making. It can, however, be called into question whether this is the best way to deal with the management of a scarce natural resource, at any rate if it is viewed as a source of material wealth. Innovations in fisheries management, not least the introduction of exclusive use rights, involve strong economic interests. Much depends on how exactly a fisheries management plan or a new, rights-based fisheries management system is designed; how use rights are allocated, whether and to whom they may be transferred, etc. Not surprisingly, the councils have become an arena where different antagonistic groups fight for their interests. Sometimes different states of the union are involved in this fight as well.

Economic interests are not the only ones that do battle in, with, or through the councils; environmental advocacy groups have increasingly targeted the fishing industry and fisheries management. To the most extreme of these groups fishing is an activity to be tolerated at best. Their view appears to be that the overriding purpose of fisheries management is preservation of nature as close to its pristine state as possible. For some years now one of their major goals has been to set aside large swaths of the ocean as protected areas where no fishing whatsoever is allowed. In the fisheries off Alaska they have succeeded in having the fishing fleet banished from areas extending as much as 10 miles from the coast so as not to interfere with the foraging of Steller's sea lions, an animal claimed to be endangered, but of no use to humans and if anything a competitor for valuable fish resources.[2]

Surf Clams and Ocean Quahogs[3]

The first fishery in which ITQs were implemented in the United States was the fishery for surf clams and ocean quahogs off the Mid-Atlantic coast and New England. The surf clam fishery was also the first fishery in federal waters to be put under limited entry.[4] This happened in 1977, immediately after the Fishery Conservation and Management Act had taken effect. The capacity of the fishing fleet had already by that time grown too large for the total catch being deemed advisable. The surf clam fishery was regulated with a 10 year horizon of secure supplies. The ocean quahog fishery developed as a substitute for the surf clam fishery after the entry to the latter had been closed. This fishery was similarly regulated for a secure supply but with a longer time horizon, 30 years instead of 10.

Despite the ban on entry of new vessels, the capacity of the surf clam fleet continued to grow. The rules for replacement of decommissioned vessels were liberal, allowing them to be replaced with larger ones; over the period 1983–1987 the fleet tonnage increased by 30 percent. The fishing capacity of the fleet also increased through technological improvements; some vessels were equipped with two dredges instead of one, the width of the dredges was increased, and larger hoses were used to flush the clams out of the seabed mud. This capacity increase gave rise to a nightmare of detailed regulations of when each vessel was allowed to fish, partly to keep the catch within stipulated limits and partly to ensure an even flow of raw material to the processing industry. These restrictions got tighter and tighter, and immediately before the introduction of the ITQs the regulation had reached a point where each vessel could fish for 6 hours every third week. This story puts in focus the basic weakness of effort control; under that regime the boatowners have an incentive to increase the fishing power of their vessels as much as the rules allow. In these fisheries the controls could however have been much more effective if the replacement rules for vessels had been tighter and a limit put on the equipment used on each vessel (such as the number of dredges).

Already at the time when limited entry was put in place it was realized that this was not a long-term solution from the point of view of efficiency. Discussions began immediately about what to replace it with. Early on individual vessel quotas came up as an option and remained the most serious option throughout, but the discussion on how to implement them was to take no less than 10 years. The main fault line was between independent operators and large, vertically integrated processors. The small, independent operators feared that ITQs would give processors too much market power by enabling them to stack their quotas on just a few vessels. Independent vessel owners cannot bypass processors, as all surf clams and ocean quahogs are processed before being sold to consumers. Processors relying on owner-operated vessels for supplies of fish had nothing to gain from ITQs.

As the debate on management options in the surf clam fishery evolved, it gradually became clear in which direction things were moving, and industry operators began planning for the contingency that individual vessel quotas would be implemented. Idle vessels were activated and fishing effort increased in order to improve the record in case quotas were to be allocated on the basis of catch history. The number of active vessels increased from 119 in 1984 to 130 in 1985 and 144 in 1986. The catch per boat rose from 20,600 bushels in 1983 to 24,900 in 1984. Positioning such as this tends to blow up the efficiency gains from the ITQs, as effort and the number of vessels subside when the ITQs have been allocated.[5] Another lesson that can be learned from this is that a protracted negotiation period before a regime change can be wasteful because of costly strategic positioning. This happened also in the Alaska salmon fishery, to be discussed below.

The final negotiations on the initial allocation of ITQs took about 2 years. For surf clams they resulted in a formula incorporating two factors, a vessel's catch history and its size, with the former being given a weight of 80 percent. Quotas were only given to vessel owners, not to crew, and the person owning a vessel when the initial allocation took place inherited the previous catch history of the vessel irrespective of previous ownership. The need to take vessel size into account arose from the replacement of old vessels with larger vessels; those who had done so felt they would be disadvantaged if the quota allocations would be based exclusively on the catch history of their previous smaller and less productive vessel. It has been alleged that this was instrumental in getting the industry to accept the allocation. In the ocean quahog fishery only catch history counted. The ITQ program was adopted by the Mid-Atlantic Council in 1989 and took effect in 1990.

In the little over 10 years since the ITQ program took effect major structural adjustments have occurred in the industry, most rapidly in the beginning. Over 2 years the fleet shrank by a half, both in numbers and in tonnage. The hours fished per year rose from 154 per vessel to 380. In 1999 there were 42 vessels in the surf clam fishery; 10 years earlier, there were 141. With fewer vessels, despite the fact that the remaining vessels are now used full time, there has been a 30 percent reduction in the number of jobs. This has weakened the bargaining power of labor, which has seen its share of the catch value decline. The cost of leasing quotas is deducted from the catch value as an operating cost before calculating the share of labor. Some operators who own quota shares have rented out their own quotas and based their operations on quotas leased from others, in order to legitimize claims to deduct leasing costs, a phenomenon that also has occurred in Iceland and caused much controversy.

Some processors have divested themselves of vessels, instead holding quota shares for ensuring supplies and letting independent operators do the fishing. The number of quota holders has apparently not changed much since the ITQs were put in place while the number of vessel owners has fallen from a little less than 60 to 21. While some have sold out of the fishery altogether, most apparently find it to their advantage to keep their quotas and rent them out on an annual basis. Thus there apparently is substantial absentee ownership of quotas. This is also a sign of the windfall that is typically cashed in by the first generation of quota owners, who are lucky enough to get quotas for free. Those who come into the industry later have to buy their way in, and the best they can hope for is to recover the money they paid for getting in and perhaps a little more, but the scope for windfall gains gets less as the restructuring of an overdimensioned industry proceeds. A familiar resentment among young crewmembers on clam vessels today is the difficulty to work one's way up and become a boatowner and a quota owner.

Like in the Alaska halibut fishery (to be discussed below), opponents of the ITQs in the surf clam and ocean quahog fishery took their case to court. Some arguments appear far fetched, such as having kept logbooks in the ocean quahog fishery with sloppiness because catch history was not expected to create rights to quota allocations. None of those lawsuits has succeeded, however. Also similar to other ITQ fisheries in the United States, the surf clam and ocean quahog quotas are explicitly defined as a revocable privilege. This is a bit surprising, since the United States is known for stronger private rights to some natural resources than obtain in most other countries; mineral rights, for example, belong to the owner of the surface but can be leased separately from the surface. Making ITQs less than full-fledged private property has had the effect that they cannot be used as collateral for loans, which makes it more difficult for newcomers to enter the fishery. In the surf clam and ocean quahog fishery this has meant that banks assume the formal ownership of quotas until the loan they secure has been paid, an avenue closed to the Pacific halibut fishery, as will become clear below.

Of all the quota systems and similar arrangements in the United States, the surf clam and ocean quahog ITQs have the fewest constraints on ownership and transferability. Anyone can own quotas in these fisheries; the quotas are not tied to specific vessels, so any vessel can fish for clams and ocean quahogs, provided it has an ITQ coverage. Furthermore, there is no upper limit on how large a share of the total quota can be owned by one firm. There is substantial concentration of ownership; ten quota owners (20 percent of the total) own 70–80 percent of the quota shares. Such concentration may just as well be desirable as the opposite. Its detrimental effects are likely to be associated with power in the labor market, but the eastern seaboard of the United States is not exactly characterized by lack of employment alternatives or long travel distances to a new job. Clam products are sold in fierce competition with other foodstuffs, so even a single producer of clams would not have much power in the consumer market. Furthermore, there may be economies of scale in this industry.

Alaska Halibut and Sablefish

The story of the Pacific halibut fishery is in some ways similar to the surf clam story. This fishery was long plagued by much too many boats chasing too few fish. Management of the Pacific halibut fishery by a limit on the total catch far predates the 200-mile limit, aided undoubtedly by the fact that only two countries, Canada and the United States, were engaged in this fishery. In 1923 Canada and the United States concluded a treaty for a joint management of the Pacific halibut fishery and set up the Pacific Halibut Commission as a technical body for this purpose. The halibut occurs on the continental shelf all the way from the Aleutian Islands down to California and is regarded as one single stock. Only an insignificant amount is caught, however, off the coast of the contiguous states; practically all the US catch is taken off Alaska.

In biological terms the management of this fishery was for a long time quite successful. Catches went from about 50 million pounds in the 1930s to 70 million pounds in the mid 1960s.[6] After that came a steep decline to a trough of about 30 million pounds in 1974, followed by a recovery back to 70 million pounds again in 1988.

Already in the early 1960s, two economists, James Crutchfield and Arnold Zellner, pointed out that, in economic terms, the Pacific halibut fishery was a failure.[7] There were no limits on entry into the fishery. The fact that the stock was being maintained at a reasonable level prevented the catch per unit of effort from falling to a level where the fishery would no longer be profitable, so more boats were attracted to the fishery. The result was that the fishing season became shorter and shorter, with more numerous and powerful boats competing for a given annual quota for fish. The fishery would be opened on a certain date and then closed when the total catch quota was estimated to have been taken. In the early 1970s the fishing season in the central Gulf of Alaska, where about half of the US halibut catch is taken, was more than a hundred days, but in the years just before the ITQ program was put in place in 1995 it was down to 2–3 days.[8]

The disadvantages of such short fishing seasons are obvious. Fishermen go to sea in hazardous weather in order not to miss that year's opportunity. The landings occur over a short period of time, and the fish are caught in a frenzy. This is likely to mean a loss of product quality; fishermen are more concerned about getting as much as possible than with maximizing the value that can be obtained from each individual fish. In the halibut fishery this meant that most of the catch was frozen instead of being sold fresh to consumers. Not only is the frozen product inferior and sells at a lower price, it also is more costly to process and store.

In 1987 the North Pacific Fisheries Management Council asked for an investigation of alternative management solutions for the Alaska halibut and sablefish fisheries. ITQs were among the options and quickly became the preferred one. There was fierce resistance, however, against the idea. The opposition had various causes.[9] One was ideology. Some people objected to privatizing a public resource through exclusive, individual use rights. A variant of that argument, but one which did not deny the efficacy of use rights for economic rationalization, was objection to allocating such rights free of charge and in addition paying for the management of the fishery with public money. At the time, cost recovery was explicitly prohibited by the Magnuson Act, but this has now been changed, and a cost recovery program has been in place in the Alaska halibut and sablefish fisheries since 2000. In 2000 and 2001 the management costs amounted to about 2 percent of the value of the catch.[10] These are the costs of the ITQ program only; the costs of the Pacific Halibut Commission are covered by the Canadian and American governments, and other costs of halibut and sablefish management incurred by the National Marine Fisheries Service or the North Pacific Management Council are covered by the federal government of the United States. Such costs would, however, be difficult to separate from other activities of these two entities.

The main opposition against the ITQ program appears to have come from Alaska while the industry based in Seattle and surroundings appears to have been more appreciative. Some Alaskans feared that Alaskan fishing ports would be disadvantaged through fish quotas being sold to fishermen from Washington State. For that reason presumably, the State of Alaska opposed the program. Then there were arguments about culture and lifestyle. Some, and perhaps major structural changes could be expected to follow from transferability of quotas, since there were much too many boats participating in the fishery. Such changes might mean that people would have to move to other places to find work. A rationalization of the fishery could be expected to lead to fewer boats and fewer fishermen, but a longer period of employment for those who remained in the fishery. All these arguments ultimately led to severe constraints on the transferability of quotas between groups of fishing vessels and between different fishing areas being built into the program; one administrator involved with the Alaska halibut fishery has called it "infernally complicated."[11] The purpose of these restrictions is partly limitation of the scope of structural changes and partly ensuring that the industry continues to consist of owner-operated boats.

The objection to absentee ownership of fish quotas is related to the lifestyle argument. Arguments such as "fishermen should not become corporate employees" were frequently heard. Others objected to the possibility that some fishermen might retire from the industry, rent out their quotas and receive "mail-box income." This resulted in the program requiring that the quota holder be on board the vessel while the quota is being fished. This has not been without problems. Some fishing operations are organized as corporations which were, because of their track record, entitled to quotas. Such corporations are allowed to fish their quota as long as they remain in the hands of the original owners, but with a change of ownership the corporation must be replaced by individual ownership. Presumably this causes some disadvantage, as the corporative form would not otherwise have been selected over personal ownership.

Much disagreement developed over the initial allocation of quotas. The Council opted for giving quotas only to owners of fishing vessels but not to crewmembers. There are undoubtedly practical reasons for this, as some crewmembers are employed only for a short period and may be difficult to track down once they have left. Controversies would surely have arisen over the question of how long a track record would be required for being eligible for a quota share. The North Pacific Management Council defended its stand by arguing that investing in a fishing vessel showed a greater commitment, *inter alia* in the form of risk taking. Some people would argue, however, that risking one's life and limb is a respectable enough risk taking; fishing in the waters off Alaska is one of the most dangerous occupations in the United States. The halibut fishery is somewhat special in that the vessel owners are also active fishermen, which makes the greater risk taking argument more valid than if the boatowners were occupiers of offices on land. A related concern was that the ordinary progression from a deckhand to a boatowner would be impeded. To some extent this is addressed by restricting transfers of quotas to those who initially received quotas or have been crewmembers for at least 150 days. At the end of 2001 quotas were held by 975 crewmembers, and they held about 20 percent of the quotas in the halibut fishery and 11 percent in the sablefish fishery.[12]

The qualifying period and catch levels were also contentious. The Council opted for 1988–1990 as qualifying years. There were good reasons for not extending the qualifying years beyond the point when the ITQ rules were framed (the first proposal was published in the *Federal Register* in 1992 to solicit public comment). This would have greatly intensified what was already a derby fishery, as people would have positioned themselves for the largest possible quota share. Two problems with only three qualifying years were that some people had just entered the fishery at the end of that period or later, and that some established halibut fishermen did not fish all three years, electing to stay out of the fishery in one or more of these years because of the very short openings. To address the latter concern, eligible boatowners were allowed to use catches from the best 5 years in the period 1984–1990 as qualifying catches on which to base the allocation of quotas.

The IFQ (individual fish quota) program, as it came to be called in the Alaska halibut and sablefish fisheries, finally made it into the *Federal Register* in November 1993 and took effect in early 1995. The consequences were strong and immediate; the halibut fishing season became extended to about eight months, and the price of raw fish increased substantially because most of the fish were now sold to the fresh fish consumer market. The number of boats participating has decreased from 3,450 (halibut) and 1,191 (sablefish) in 1994, immediately before the program took effect, to 1,451 (halibut) and 433 (sablefish) in 2001.[13] One group which has lost out is previous crewmembers. The membership of the crewmembers' union in the Seattle-based longline fleet has been about halved since the program came into being. In early 2002 members of the union spoke bitterly of having been left out in the cold by the boatowners who, in return for the union's support of the IFQ program in the early 1990s, offered an eight year long contract, up for renewal in 2002. The boatowners wanted a hefty rise in their share of the catch (from 31 percent to 37–40 percent), and the union realized that its bargaining position had gravely deteriorated. Fewer fishermen were in demand, and the more leisurely pace at which fishing could be conducted made it less dependent on skills useful for operations where time and swiftness are extremely important.

The attitudes among the Alaska halibut fishermen toward the IFQ program was so negative on the eve of its implementation that it is somewhat puzzling that it ever took effect. According to a survey of boat captains in 1994 most seemed unsupportive or outright hostile.[14] Almost 70 percent of the respondents thought the allocation of quotas was unfair, and only 30 percent supported the program over and above other solutions, even if it came highest among the several alternatives offered. About 40 percent expected to become financially worse off because of the program, and only 25 percent expected to become better off.

After the program had started to work some previous opponents on the boatowners' side changed their mind, but up to and beyond that time opposition was fierce. The opposition took its case to court, and a final verdict was pronounced in the Ninth Circuit Court in 1996. The question at hand was whether the decision to implement the quota program was "arbitrary and capricious." The court found it was not but apparently could not resist a lengthy exposition which for a time was cited with great relish by opponents of the IFQ program:

This is a troubling case. Perfectly innocent people going about their legitimate business in a productive industry have suffered great economic harm because the federal regulatory scheme changed. Alternative schemes can easily be imagined. The old way could have been left in place, where whoever caught the fish first, kept them, and seasons were shortened to allow enough fish to escape and reproduce. Allocation of quota shares could have been made on a more current basis, so that fishermen in 1996 would not have their income based upon the fish they had caught before 1991. . . .[15]

The court was apparently not troubled by the inefficiency of open access. Of course the derby style could have been left in place, the fishing season perhaps having to be shortened to hours instead of days. The qualifying period could have been extended right up to the beginning of the program, to produce a still greater frenzy as people jockeyed for position, further shortening the fishing season. This formulation betrays either a total disregard for economic efficiency or a high degree of economic illiteracy.

Wreckfish[16]

The wreckfish fishery is an interesting exception from other ITQ fisheries in the United States in that the ITQs were put in place before a situation of gross overcapacity had developed and in response to an initiative from the industry itself. The wreckfish is a grouper-like fish located at great depth (several hundred meters) about 120 miles off the southeastern United States. This fishery is both recent and small. It began with two vessels landing less than 30,000 pounds in 1987. But it developed quickly. In 1989, more than 2 million pounds were landed by 25 vessels; in 1990, 4 million pounds by 40 vessels. In 1991 a catch limit of 2 million pounds was put into effect. No less than 90 vessels competed for this catch.

The pioneers in this fishery quickly became concerned that its profitability would be eroded by the new entrants who kept on coming in. They persuaded the South Atlantic Fisheries Management Council to implement an ITQ regime in 1992 and allocate the quotas to those who had been active in this fishery. To be eligible one had to have caught at least 5,000 pounds in 1989 or 1990. Half of the quotas were allocated on the basis of catch history, and the other half as equal shares to all. No one could get more than 10 percent of the initial allocation. In the initial allocation, in April 1992, 49 boatowners received shares. There was immediate consolidation; by mid August there were 37 left, by June 1993 there were 31, and by May 1994 there were 26.

This case is in some ways similar to the producers' cooperatives on the Pacific coast of the United States, to be discussed below, where the industry has also taken the initiative to establish exclusive use rights, in line with what the economic theory of property rights would predict. The wreckfish fishermen were undoubtedly helped by the fact that they were not very many. They apparently found a receptive ear among fishery administrators in Charleston who seized the opportunity to put ITQs in place in a "greenfield" situation, a bit similar to the situation in the deep-sea fisheries of New Zealand, which was discussed in chapter 5.

The ITQ regime notwithstanding, the fishery for wreckfish has declined. In the mid 1990s landings were only about 25 percent of the total quota, and since 1998 they have been insignificant. Apparently the wreckfish fishery is not profitable enough to attract even those who hold an exclusive right to participate in it. The apparently high profitability in the late 1980s and the early 1990s may have been associated with fishing down a virgin stock and thus not sustainable. The boats participating in the wreckfish fishery were also active in other fisheries and could easily switch out of wreckfish in case that fishery was not profitable enough.

The Alaska Pollock Fishery and the ITQ Moratorium

What occurred in the ITQ fisheries in the first half of the 1990s strengthened the forces both for and against putting other fisheries in the United States under ITQ regimes. To many these developments seemed a resounding success. The industry became more efficient, fishing effort was reduced, the fishing season became longer, and the fish was turned into a more valuable product. The time seemed ripe for applying this regime to other fisheries in the United States. Others saw undesirable consequences, or regarded as negative effects that others viewed as positive. Concentration of quotas in fewer hands, higher price of quotas and barriers to entry, more supplies into the fresh fish market and less processing, decline in employment of fishermen and their wages; these were viewed by some people as undesirable and not to be repeated in other settings.

One fishery which many by the mid 1990s expected would be put under an ITQ regime was the Alaska pollock fishery. As always happens in such cases, individuals with foresight tried to position themselves advantageously. A consortium of individuals in Seattle bought a sunken trawler, hoping that its catch record would entitle them to an allocation of ITQs. An ITQ regime appeared both overdue and particularly suitable for the Alaska pollock fishery, which is relatively new. Alaska pollock is a bit like cod and Atlantic pollock (alias saithe) and is found in large quantities in the Bering Sea and the Gulf of Alaska. This fishery was almost insignificant in the 1950s but grew rapidly in the 1960s and early 1970s. In the beginning it was mainly conducted by Japanese vessels which turned the fish into surimi, but when the 200-mile limit was established in 1977 most of the fishing grounds came within the economic zones of the United States and the Soviet Union. The catches fell for a few years as the foreign fleets were barred from the exclusive economic zones, but started growing again as the United States entered into joint venture agreements with foreign vessel owners and developed its own fleet.

The United States controls its pollock fishery through a total catch quota. The catches in this fishery have been remarkably stable since their initial growth in the 1980s, as shown in figure 7.1, while the catches of other nations have peaked and declined. Some would see the US Alaska pollock fishery as a model of sustainable fisheries; despite an undoubtedly variable environment the total catches have been kept stable, possibly at the

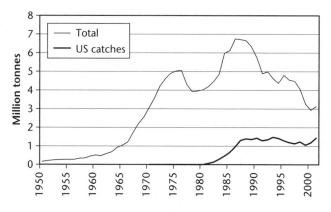

Figure 7.1
Catches of Alaska pollock. Source: fishery statistical database, UN Food and Agricultural Organization.

expense of forgoing opportunities of catching more during productive periods. Furthermore, the Alaska pollock trawl fishery is probably one of the cleanest fisheries in the world, in the sense of taking little of other species than the single one targeted.

Until the early 1990s there was free competition for the total quota and, as expected, the fishing season shrank as more and more vessels entered the fishery. An investigation of the fishery in the latter part of the 1980s concluded that the fleet participating in this fishery was already too large,[17] but nobody particularly wanted to listen to that message. Everybody was busy making deals, thinking he would win the race for the fish. As the fleet continued to grow and the fishing season became shorter the pressure mounted to do something about this. In 1991 the total catch quota was for the first time divided between the inshore and offshore fleets. The inshore fleet consists of boats delivering their catch to shore-based plants in Alaska while the offshore fleet consists of factory trawlers, motherships that process but do not catch fish, and catcher boats delivering fish to the processing vessels.

This division of the total catch proved contentious. A struggle developed over the spoils of this highly profitable fishery. The inshore-offshore allocation was changed twice, in 1995 and again in 1998. Furthermore, a part of the total catch quota was set aside for the development of certain coastal communities in western Alaska. These communities had little or no history in the fishery for pollock and few of them utilize these quotas on their own, instead entering into joint ventures with fishing firms outside these communities, some from Washington State. After 1995, 65 percent of the total catch was assigned to the offshore sector and 35 to the inshore sector, after subtracting 7.5 percent for the Community Development Quota) and a smaller quantity to cover bycatch in other fisheries.

The conflict between the offshore and inshore sectors over the distribution of the total catch was to have an important bearing on the plans for ITQs in the Alaska pollock fishery. Another conflict of a similar importance emerged between the catching sector of the industry and the shore-based processing plants. Fishing the total quota over a short period made it necessary to build up a processing capacity which could cope with landing large quantities of fish over a short time period, but with ITQs and a longer fishing season some of this processing capacity would not be necessary. With ITQs, processors would therefore probably compete with one another for fish for their plants, which would raise the price of fish and possibly put some of them out of business. Giving ITQs to boatowners would thus benefit them in two ways; by giving them a valuable asset and by raising their revenues at the processing industry's expense. Some economists proposed a "two-pie" solution to deal with the possible losses of processors, by which individual processing quotas would be given to the processors, guaranteeing them fish supplies and eliminating the element of competition for the fish.[18]

Many economists are critical of the two-pie system, because this is not needed to deal with the wastefulness of the race for the fish. With a given total quota, competition for fish in the marketplace only affects the division of income between fishermen and processors. Overcapacity in the processing industry is one of the undesirable side effects of open access to the fish stocks, and when open access is done away with this overcapacity will have to disappear. Bankruptcies are one method, and if deemed too harsh and undesirable, there are other methods such as buy-outs of overcapacity which can be employed. The processing industry's stand against ITQs in the pollock fishery is a good example of how the waste accompanying open access generates vested interests and constituencies which will work to perpetuate this system and hinder efficiency-promoting reforms such as ITQs. Gains in the form of the disappearance of excessive processing equipment and the shedding of redundant labor will translate into pecuniary losses of the individuals involved, and they will therefore oppose such changes unless adequately compensated.

In 1996 the Magnuson Act, on which the regulations of the American fisheries in federal waters are based, came up for review in Congress. This revision developed into a major struggle between adherents and opponents of ITQs. Some environmentalist organizations, Greenpeace in particular, were heavily engaged in opposing ITQs and lobbying Congressmen and -women for banning the use of ITQs in American fisheries. The environmentalist opposition appears largely ideological; it opposes profit-driven fisheries, large-scale operations, concentration of ownership of fishing rights, and suchlike. There are, however, environmentalist organizations, such as the Environmental Defense Fund, which support ITQs as market-based tools to cope with overfishing and overcapacity in the fishing industry.

In any case, a decisive alliance developed between the environmentalists and influential senators. As the Congress was debating the Magnuson Act, an ITQ plan for the red snapper fishery in the Gulf of Mexico, having been thrashed out by the Gulf Fisheries Management Council, was waiting for the approval of the Secretary of Commerce. Those parts of the industry opposing ITQs found allies in Senators John Breaux of Louisiana and Trent Lott of Mississippi. And the Alaska-based processing industry found a powerful ally in Ted Stevens, a senator from Alaska. The inshore-offshore conflict, already mentioned, is a geographical one between the state of Alaska and the state of Washington where the offshore processors and associated catcher boats are based. This conflict is in fact reminiscent of the conflicts between coastal states and distant water fishing nations over jurisdiction at sea, even if in this case it occurs within the jurisdiction of a single, federal state. The coastal communities of Alaska can be compared with Iceland and Northern Norway; in all of these places fishing is the single most important industry, and distance from major markets and other disadvantages hinder the development of other economic activities. The Bering Sea is farther from Seattle than Iceland is from the Humberside, where the English offshore trawlers were based.

Not surprisingly, Senator Stevens cared more for the processing industry in Alaska than the Seattle-based fleet. He also had a further agenda; to eliminate the boats, mainly Norwegian owned, which had taken advantage of a loophole in the Anti-Reflagging Act of 1987, a protectionist legislation banning the use of foreign-built fishing vessels.[19] The controversial vessels had been rebuilt in Norway, retaining little other than the original American keel. Paradoxically, the shore-based processing industry in Alaska is to a large extent Japanese owned.

The debates over the reauthorization of the Magnuson Act, now known as the Magnuson-Stevens Act, resulted in a four-year moratorium on the use of ITQs in the US fisheries. The moratorium was to a large extent due to Senator Stevens's agenda. ITQs in the Alaska pollock fishery would have grandfathered in the Norwegian-owned vessels as quota holders, and free trade in quotas might very well have strengthened the Seattle-based fleet relative to the Alaska-based fleet and processors.

Undoubtedly the ITQ moratorium delayed and distorted the development toward private use rights in the American fisheries. The already mentioned ITQ program for the red snapper fisheries in the Gulf of Mexico was killed by it while sitting on the desk of the Secretary of Commerce. One ostentatious argument for the moratorium was that ITQs needed further study. This study was commissioned from the National Research Council, and the committee put together for this purpose spent considerable time on travels, hearings and deliberations in various parts of the United States, also taking into consideration the experience in some other countries where ITQs have been implemented. These efforts ended in a study published in 1999 which included a list of recommendations, the very first of these recommending that the moratorium should be called off and the use of ITQs permitted where appropriate.[20] That notwithstanding, when it would have otherwise lapsed in 2000, Congress extended the moratorium for 2 more years. The moratorium finally lapsed in October 2002.

The Whiting Cooperative

The ITQ moratorium did not eliminate the incentives to work toward exclusive use rights, it only channeled them into new directions. The potential gains from closing the access to profitable fisheries were still there, and the problems accompanying derby-type fishing had not gone away. While in other cases the moratorium stalled or derailed the development toward exclusive, individual rights, they took on a new guise in the Alaska pollock fishery, the so-called fishing cooperatives.

The development toward fishing cooperatives originated in the offshore trawl fishery for Pacific whiting off the coasts of Oregon and Washington, before the ITQ moratorium. The total catch quota for Pacific whiting is divided between the offshore and the inshore fleets, as in the Alaska pollock fishery. There are only a handful of firms engaged in the offshore fishery for Pacific whiting, and the leaders of those firms began to talk among themselves about ways to end the mutually disadvantageous derby-type competition. With the inshore-offshore allocation having been made, and provided it would not be changed whimsically, it would make sense for the industry to be authorized to administer its own quota for its own benefit, rather than waste its efforts on competing for a given quota and shortening the fishing season. These ideas came about partly out of fear that the Pacific Management Council would not be able to deal with the issue of allocating ITQs in a timely fashion. Events were unfolding in the Pacific halibut IFQ program and demonstrated the difficulties the North Pacific Council had in dealing with this. (About 7 years passed between the time ITQs were first proposed for these fisheries and 1995, when they took effect.) These industry players believed they could deal more effectively with the allocation of quotas, to their mutual satisfaction.

These musings resulted in an initiative to form a producers' cooperative. The cooperative essentially amounts to an ITQ management system, albeit with certain restrictions and imperfections compared to an idealized system. The purpose of the cooperative is deciding how its share of the total permitted catch is to be taken. In practice the members of the cooperative have allocated their share of the catch among themselves on the basis of recent catch history, although there has been some give and take among the members. With the amount of fish to be caught by each producer being given, the producer has every incentive to maximize the value of his quota by minimizing cost, selecting the most profitable product mix, and improving product recovery rates. By trading quotas among themselves, the producers may be able to raise the profitability of their operations still further.

Before the cooperative could be put in place, it had to be clarified whether or not this would violate the antitrust law of the United States. The law explicitly forbids cooperation among producers for price fixing and market sharing but such, needless to say, is not the purpose of the cooperative. The initiators of the Pacific whiting cooperative contacted the Department of Justice about the legality of their plans. After the Department of Justice had confirmed that the planned cooperative did not violate antitrust legislation the cooperative was formed. Agreement on how to share the catch allocation was apparently swift, reached in a matter of hours, aided undoubtedly by the fact that there are only four companies in this fishery. Because the Department of Justice gave its opinion about half way through the 1997 season the fishery started out as a competitive race for a given total quota. The vessel quotas were introduced immediately after the opinion of the Department of Justice became available and resulted in substantial gains through a reduction in the number of vessels fishing, improved product recovery rates, and less bycatch of unwanted and unauthorized fish.[21]

The Alaska Pollock Fishing Cooperatives: ITQs through the Back Door

The positive experiences with the whiting cooperative and the fact that ITQs had been put on hold stimulated the interest in a similar arrangement for the Alaska pollock fishery; the companies which are engaged in the Pacific whiting fishery are also involved in the Alaska pollock fishery. The Alaska Pollock fishery is, however, a considerably more complicated case, being troubled by conflicts between the offshore and inshore sectors, and the catching sector and the shore-based processing plants, as already explained. But as political dealings had foreclosed the ITQ option, political dealings could be employed to redress the balance. In 1998—2 years into the original ITQ moratorium—parts of the Alaska pollock industry had successfully lobbied for a legislation that realized many of the gains an ITQ system would have brought. The result was the American Fisheries Act, attached as a rider to the Senate appropriations bill (Senator Stevens was at that time chairman of the Appropriations Committee). This is a remarkable piece of legislation, and its smooth passage in the form of a rider to the appropriations bill stands in a marked contrast to vigorous debates in Congress on ITQs, their merits and demerits, and the ensuing moratorium. The American Fisheries Act brought ITQs back through the back door, albeit in a somewhat different form, through authorizing the establishment of producers' cooperatives in the Alaska pollock fishery. As a prerequisite, it specified the division of the total catch quota for Alaska pollock in the Bering Sea–Aleutian Islands fishery between three vessel groups; motherships and associated boats, factory trawlers and associated boats, and the land-based processing industry in Alaska and associated boats.[22]

Furthermore, the Act substantially changed the allocation of the total catch between the inshore and offshore sectors of the Alaska pollock fishery. The 35–65 division of the "directed fishery" (i.e., what remains after the Community Development Quota and the bycatch allowance) was changed to an even split, substantially favoring the inshore sector. Distinction was made between two sectors of the offshore fishery, with 10 percentage points allocated to the motherships and 40 percentage points to the factory trawlers and associated catcher boats. The Act also raised the Community Development Quota to 10 percent of the total allowable catch.

The Act went a long way toward realizing the ideas behind the two-pie system, i.e., preventing mutually disadvantageous competition among Alaska-based processors. To this effect, any producers' cooperative in the inshore sector has to be formed around one particular processor. A boat is qualified only for a cooperative delivering to the processor to which it delivered the plurality of its pollock catches in the previous year, and any cooperative has to deliver 90 percent of its members' catch to its processor.

Furthermore, the Act prevented access of new vessels to the pollock fishery. In the inshore sector this was done by reserving the right to continue in this fishery for vessels delivering more than a certain minimum of pollock in 1995–1997. In the offshore sector the Act explicitly named the vessels eligible to continue fishing. The excluded vessels, also explicitly named in the Act, were bought out of the fishery for $95 million, $20 million of which was a federal grant and the rest a federal loan payable through a fee on the inshore fisheries. This can be viewed as the Alaska-based sector's payment for its increased share of the total catch quota; the catches of the vessels bought out of the offshore fishery were approximately equal to the quota transferred to the inshore fishery.[23]

After the American Fisheries Act had been passed, producers' cooperatives were soon formed in all three sectors of the Alaska pollock fishery. In the mothership sector there is one cooperative. In the so-called catcher/processor sector (factory trawlers and boats that deliver fish to these) there are two cooperatives, the High Seas Catchers' Cooperative, which includes the catcher boats delivering their catches to the factory trawlers, and the Pollock Conservation Cooperative, which includes the factory trawlers. The Pollock Conservation Cooperative receives the bulk of the allocation to this sector, or 36.6 percentage points of the 40 allocated to the sector, with the remaining 3.4 percentage points going to the seven catcher boats. The two cooperatives cooperate on monitoring and planning catches.

According to the 2001 joint report of the two cooperatives, all seven catcher boats leased all of their catch allocations to the factory trawlers, and in the two previous years this was true for most of their quotas. These boats have thus been effectively removed from the fishery; instead they participate in other fisheries and get a share of the rent in the Alaska pollock fishery through leasing their quota. Since the factory trawlers no longer need to compete for a given overall quota they have been able to catch the fish that the catcher boats used to take. Needless to say, these gains could be largely or wholly eroded by aggravating overfishing of stocks to which the catcher boats leasing their quotas turn. This is dealt with through so-called sideboard restrictions, regulating how much a pollock vessel can catch in other fisheries.

The factory trawlers' cooperative has resulted is a more efficient utilization of these vessels and a longer and more leisurely fishing season; the season doubled from 75 days in 1998 to 149 in 1999, the first year of the cooperative.[24] Another manifestation of the cost savings due to the cooperative is that the top weekly landings in the two fishing seasons in 2000 were about half of what they were in 1997 and the seasons were correspondingly longer, the total catch being about the same in both years. On the revenue side this has resulted in higher product recovery rates, as shown in table 7.1. Note that the factory trawlers' cooperative was formed in 1999 while the motherships' cooperative was formed a year later. In both cases there was a significant increase in the product recovery rate after the cooperatives were formed. There is less increase in product recovery rate to be noted for the inshore processing plants, which is not surprising; the reason given for the increase in the at-sea product recovery rate is that fishing can now proceed at a more leisurely pace, providing time to better tune filleting machines and other equipment to the fish they were handling.[25]

The Pollock Conservation Cooperative has a research program funded through a voluntary levy on the catch. In 2000 this amounted to $1.4 million, allocated between three universities, with the University of Alaska getting the lion's share.[26] This can perhaps be taken as a sign of a "stewardship effect," i.e., the members of the cooperative realize that the future

Table 7.1

Product recovery rates in the Bering Sea/Aleutian Islands pollock fishery. (The cooperative in the factory trawler sector was formed in 1999, the others in 2000.) Source: NOAA/NMFS 2002, pp. 4–35.

	1998	1999	2000
Factory trawlers	20.3%	25.5%	27.5%
Motherships	20.7%	20.8%	26.8%
Inshore processors	35.7%	35.8%	36.6%

profitability of their operations depend on how well the fish stocks are managed and are ready jointly to spend money on research for improving the management.

In the Alaska-based processing sector seven cooperatives have been formed. As previously stated, in this sector cooperatives must be formed around a particular processor. The rules are complicated, being the result of a fight over the rent between catchers and processors. The raw fish price determines the share going to each, so giving undue market power to either side needs to be avoided if the rent is to be shared equitably. There is some presumption that the processors will lose market power if the fishing season is stretched.[27] Two factors mitigate this. First, there are only six processing companies in the inshore fishery, so a cutthroat competition among them is somewhat unlikely. Second, most of these companies have boats of their own. This makes them better informed about the costs of fishing than the owners of the catcher boats are about processing costs.[28]

Briefly, the rules on forming cooperatives in the onshore processing sector are as follows. The designated processor must agree to process the fish caught by the boats in the cooperative. The owners of at least 80 percent of all the boats delivering to the processor must agree to form a cooperative. This is an unusually stiff qualification requirement. The rationale for it lies in the processors' interests; this way they are assured of a large share of the catches taken by the members of the cooperative.

The rule that any cooperative must deliver at least 90 percent of its members' catch to its processor is obviously designed to strengthen the market power of the processor; the members of the cooperative are free to scout around for the best price for only 10 percent of their catches. To further enhance the market power of the processors there is a requirement that a member who leaves a cooperative must take his catch share to the common pool for one year and compete with others before joining another cooperative. The 10 percent rule provides a loophole, however; members of a cooperative can stack this on one or just a few boats and have them deliver more of their catch to another cooperative and become eligible for membership in that cooperative next year. Five boats switched between cooperatives between 2000 and 2002 without spending a year in open access.[29]

Lessons from the Story of the Producers' Cooperatives

The story of the producers' cooperatives in the Pacific whiting and the Alaska pollock fisheries shows that under the right circumstances the fishing industry is capable of dealing effectively with open access, replacing it with internally negotiated arrangements that reduce the economic waste accompanying this regime. Certain preconditions must, however, be met. A limit must be set to the overall catch, and entry of new boats has to be precluded unless their owners buy quotas from somebody already in the fishery.[30] The legality of trading quotas between boats or companies must also be assured. Under those conditions, the rents to be realized from cost cutting and better utilization of catches act as a driving force to bring about greater efficiency. For this to be possible, at least some of the rents will have to accrue to the industry; if that did not happen the industry would have no incentive to rationalize its operations. What scope there is for rent capture by government under this scheme of things is unclear. A solution where rationalization is enforced from above, perhaps through auctioning of quotas, is another story, but that would be the outcome of a government initiative, not an industry initiative.

A solution like the US Pacific fishing cooperatives has several drawbacks compared with an ideal ITQ system, even if it works like ITQs. Splitting up the total catch between fleet groups and prohibiting trading of quotas between them precludes the efficiency gains that might lie in reducing the number of vessels in less efficient fleet groups or phasing them out altogether. More generally, any restrictions on trading quotas, such as tying inshore fishing boats to a certain fish processor, mean potential economic losses. If the quota allocations are tied to boats and boatowners are not permitted to sell quotas permanently they will hang on to their boats, spending resources unnecessarily on maintaining them or using them if that is required. Uncertainty about whether the arrangement will last will also similarly entice boatowners to maintain their boats. Originally the

American Fisheries Act had a sunset clause limiting it to 6 years, but this clause was removed in a rider to the Senate Appropriations bill passed in 2001. Both leasing of quotas and permanent buying and selling is now taking place, both in the inshore and offshore fishery. Informal price quotations as of the summer of 2003 were $1,200 per tonne in the inshore sector and $2,500 for the offshore sector for permanent transfers. This would indicate that the offshore fishery is about twice as profitable as the inshore fishery.

But the cooperatives may have some advantages compared with ITQ systems without any restrictions. ITQs are demanding in terms of monitoring and enforcement. By having a small club of like-minded and similar operators assign ITQs among themselves, better compliance through peer pressure may be expected. A general drawback of ITQs is that they are use rights and not ownership rights to fish stocks; ITQs only amount to a collective and indirect ownership of the stocks being fished by the quota holders. Taking advantage of such ownership for preserving the future productivity of the stocks, through industry lobbying for cautious and small catch quotas and avoiding highgrading and discarding, is likely to be easier if the ownership is shared among few like-minded persons, especially if they have to cooperate through some formal structure like the said fishing cooperatives. Finally, producer cooperatives might be a way to cope with deep mistrust and rivalry between different groups of fishermen. If more harmonious subgroups can be identified, the total allowable catch could be divided between these, as between the onshore and offshore sectors in the Alaska pollock fishery, and the groups then left to work out the administration of the their respective quotas. A cleavage frequently occurring is one between small boats and large boats, to some extent associated with individual versus corporate ownership. In Iceland there is a special quota system for small boats, and transfers of quotas from small boats to large boats is not permitted. The management system in Norway seems to be evolving in a similar direction. In Chile a distinction is made between industrial and coastal fisheries.

The Alaska Crab Fisheries

The Alaska crab fisheries are among those in which a development toward individual property rights was delayed by the ITQ moratorium. In the early

1990s the need to rationalize these fisheries had become apparent, and many industry players favored an ITQ regime over other solutions. The moratorium, needless to say, put any such plans on hold. In the meantime the need for rationalization became more pressing, as fishing capacity increased while the crab catches dwindled. The North Pacific Management Council began developing a plan for an ITQ regime in 2001, and in the summer of 2002 it sent a letter to Congress asking for its approval of such a regime, as the ITQ moratorium was then still in effect.

As of the summer of 2003 the crab ITQ plan was still pending approval, but in anticipation that this will be forthcoming we may note its salient features. The Council has gone for the so-called two-pie system, in that not only is the allowable catch divided into ITQ holdings but the processors have also got their individual processing quotas. This is accomplished by dividing the individual catch quotas into two classes. Ninety percent of the quotas are Class A quotas, meaning that the fish must be landed to a processor having a processing quota, while the remaining Class B quotas can be landed anywhere. The purpose obviously is to avoid competition for raw crab among the processors. Politically the proposal is probably critically dependent on processors' support, which is why the somewhat cumbersome two-pie solution was integrated into the proposal.

The proposal itself in fact speaks of a three-pie system and not a two-pie one. The reason is that in addition to the fish and the processing quotas there are also regional quotas. The shares of the northern and the southern regions in the processing will reflect their shares of landings and processing in the past. This regionalization applies to the Class A quotas, or 90 percent of the total catch. The northern region is defined as the areas north of latitude 56.20'N bordering on the Bering Sea, while South is everywhere else, including, for example, Kodiak island and mainland areas bordering on the Gulf of Alaska even if north of the said latitude.

Additional features of the program are a community development quota of 10 percent of the total catch, captains' allocations of 3 percent of the total quota, and a loan program enabling captains and crews to buy quota shares. Comparing this to the established ITQ programs in the surf clam and wreckfish fisheries one can see how over time an attempt has been made to address complaints from groups fearing to lose out from an ITQ program and to avoid major structural changes in the industry. Unfortunately, the two- or three-pie solutions makes the program unnecessarily complicated and unwieldy, and they will also detract from the efficiency gains from the program by making consolidation to fewer and more efficient processing units less easy. Such consolidation would most likely mean regional concentration, which will be made more difficult by the regional quotas.

Developments and Tendencies

Looking at how fisheries management has developed in the United States since the 200-mile zone came into being, several features stand out.

First, and perhaps most important, there are the fisheries management councils set up contemporaneously with the zone. The councils were set up as public institutions characterized by openness and transparency. Their track record raises serious questions, however, about how effective it is to regulate fisheries through a political process. The councils have largely failed to deal with the problems of open access in a timely manner. It took many years to get the ITQ regime going in the surf clam and ocean quahog fisheries and the Alaska halibut fishery. It was for a long time clear what would be the fate of the Alaska pollock industry; both theory and experience tell us in no equivocal terms what will happen in a profitable open-access fishery. In the Alaska pollock fishery one could see it happen over a time span of about 10 years. After the 200-mile zone was established and the foreign fleets were displaced the American fishery got started, first through joint ventures and then with US-registered but partly foreign owned vessels. The fishery was highly profitable, the number of vessels increased year by year, but since the total catch quota did not change much the fishing season got shorter and shorter. The arguments for imposing a management system based on ITQs seemed overwhelming. This need not have meant a privatization of the value of a public resource; some of the rents in the pollock fishery could have been captured, for the interest of the general public, through fees or auctioning of quotas. Yet the North Pacific Council did nothing to rationalize the pollock fishery for many years. In the end, interference by Congress ensured that access to the fishery was closed in a way that served select special interests. Ironically, what made that interference necessary was the ban on ITQs imposed by Congress itself.

The inability of the management councils to act in a timely fashion is due to their role as mini-parliaments for fisheries where fighting among antagonistic interests takes place. Resolving these fights has proven to be lengthy and costly and has often resulted in designs that are far from ideal (the severe restrictions on transferability of quotas in the Pacific halibut fishery is a case in point). Not all of those interests are industry interests, environmental advocacy groups have also been prominent in the council process. It is no coincidence, therefore, that the industry has found a smart way to bypass the councils, going to Congress directly and have it enact laws that serve the interests of industry. Again the result is less than ideal, but more swift and maybe no worse than would have been attainable through the council process. It is possible that the council process has become so unwieldy and inert that it will increasingly be bypassed by industry, provided Congress is willing.

Better still would be to get politics out of fisheries management, replacing the council process with market-driven incentive structures that will ensure maximization of economic benefit in the long term. While phrases such as "transparency" and "stakeholders' representation" have become mantras among the chattering classes, it is highly doubtful whether this is what is needed in order to maximize the contribution of this industry to our material wealth. Not all kinds of decisions are best left to parliamentary processes and resolved through a public debate. It seems that we owe a lot of our material wealth to decision making in restricted forums, or at the individual level, where the objective is maximization of individual wealth or the wealth of a small, restricted group. It is highly likely that if the use of land, allocation of investment funds, and decisions about what to produce, where, and how, were to be decided in elected or appointed committees with a high level of transparency and stakeholders' representation the outcome would be less desirable than when these decisions are driven by profit-seeking individuals or firms, tempered and coordinated through a well-functioning market. The simplest and yet probably the best fisheries management system one could hope for is the setting of total catch quotas on the basis of expert advice (fisheries biologists and economists) and unfettered buying, selling and leasing of shares of the total catch quota.

Second, as rationalization schemes such as ITQs have developed to deal with the problems of open access, we have seen how the focus has increasingly shifted from efficiency gains to distributional issues. Various interest groups fearing that they might lose out from the implementation of such schemes have mobilized in order to protect their interests. While such mobilization is understandable and natural, it has had the deplorable effect of making the rationalization programs less functional, through various restrictions being placed on the transferability of quotas and who can own quotas and who cannot. Such catering to special interests seems particularly doubtful when it comes to excess processing capacity which has mushroomed because of short and intense seasons in derby-style fisheries. Designs such as the two-pie or three-pie systems are utterly unnecessary for regulating the fishery effectively; their only purpose is to protect processors' vested interests, but while doing so they make the regulatory system unnecessarily unwieldy. If it is deemed necessary to compensate processors there are better ways to do so, such as buying up redundant processing capacity.

Finally, over the last quarter-century environmentalists have become increasingly involved with fisheries issues, not only in the United States but also elsewhere. This has shifted the focus, both of public debate and research, away from issues having to do with wealth creation in the fisheries sector to purely environmental concerns such as biodiversity, essential fish habitat, marine protected areas, and foraging of Steller's sea lions. These developments are similar to what we find in many other areas, such as preservation of forests and keeping the Arctic Wildlife Refuge closed to oil drilling. To some this may appear a natural consequence of the United States and other industrially developed countries becoming richer and less dependent on natural resources. Others find this less than persuasive; even if fewer of us are involved in the production of food, including fishing, this is still a vital activity for our existence and well being. Nor are we likely to be able, or willing, to do without fossil fuels and timber and paper for a long time to come.

Conclusion

Privatizing fish resources can be difficult, even in countries that base their economies on private property rights and market processes. While rights-based fisheries management can be expected to bring overall gains, both to society in general and to the industry, the opposition to this type of arrangement has been formidable and sometimes decisive.

What accounts for the opposition? Even if an institutional change enhances productivity and provides net benefits to society, it typically involves structural changes that make some people worse off, unless a mechanism is put in place to compensate for their losses. There are groups within and outside the fishing industry that, in the absence of such a compensation mechanism, are likely to lose from establishing private use rights in fisheries. The gains to the industry are due to a rise in productivity, both through cost cutting and through increased value of the product. In fisheries where there is too much capital and labor, the realization of these productivity gains requires, on the cost side, that some capital and labor leave the industry, which means that capital owners go broke or sell out at a loss and fishermen lose their jobs. Those so affected are bound to lose from this process unless they are somehow compensated.

One attraction of transferable fishing rights is that they can be designed as an automatic compensatory mechanism for such changes. Fishermen who are given such rights on the basis of their previous catch history can sell them to other fishermen and leave the industry with compensation. Since they do so voluntarily, the implication is that they regard the compensation as satisfactory. That they might come to regret it later on the basis of new information is another issue; such things happen in many areas of life and are impossible and usually undesirable to guard against. People sometimes find that they could have gotten a better price for their company stock or their house if they had sold a little later, and no one should be surprised at seeing this happen with fish quotas. None of us can predict the future exactly, and who should be burdened with paying for our mistakes?

Transferable fishing rights are only a partial compensation mechanism, however, on two counts. First, the future gains from individual transferable quotas, or from other use rights such as fishing licenses, become capitalized into a market value of these rights and are mainly cashed in by those who initially got them for free; those who come into the industry later will have to buy their way in, and for them the value of the ITQ or the fishing license will be a cost that they can hope to recover when they in turn leave. This partiality across generations is not, perhaps, a serious issue; it can be argued that newcomers always have to adapt to institutions that are already in place while those who have established themselves within an old arrangement that is about to be abolished for something better should indeed receive some share in the gains from the new and better arrangement. It could also be argued that if no fishing rights were put in place the future fishermen would just cover their necessary capital costs (which would not include any payment for fishing rights), exactly as in a long-term equilibrium with fishing rights (where again they would just cover their capital costs, but this time also including the value of the fishing rights). The intergenerational equity problem would not be a lack of compensation of future fishermen but a disproportionate share of the "first" generation in the benefits of the fishing rights regime. Second, and more seriously, fishing rights such as ITQs have usually been given to boatowners only; they have seldom been given to ordinary crewmembers without any investment in the industry. When such rights have taken the form of licensing fishing vessels it is difficult to see how ordinary crewmembers could have been included, but giving them fish quotas is not in principle impossible while in practice it may be difficult. However that may be, crewmembers without the prospect of getting any quotas have little to gain and perhaps a good deal to lose from bringing in ITQs. As the demand for fishermen falls because of rationalization of the fleet, their share of the revenues generated in the industry is likely to be eroded. This has happened in the Alaska halibut fishery and in the ocean quahog and surf clam fishery on the eastern seaboard of the United States. In Iceland there have been repeated conflicts between boatowners and crewmembers, as the former have wanted to subtract expenses for leasing quotas from the revenue before sharing it with the crew. The processing industry also has little to gain and probably something to lose if fleet rationalization stretches the fishing season and there is less need for processing capacity to cope with peaks in landings. Groups that stand to lose will, unsurprisingly, use whatever political clout they may have to prevent expected losses. The pollock and crab processors in Alaska did so successfully. The fact that these losses do not represent any overall economic losses but rather reflect an economic gain is a different issue, to which I shall return.

Among groups outside the fishing industry that might lose from rights-based regimes are people living in fish-dependent communities that might be left stranded if fish quotas are sold to buyers located in other communities. Small coastal communities in Alaska feared that this would be the result of the halibut ITQ program. Some small coastal communities in Iceland have experienced this. A mitigating factor is, however, that such communities would seem to be in trouble anyway. If there are economies of scale in fishing and fish processing, the industry is likely to relocate away from small fish-dependent communities in any case. If small fish-dependent communities lose their quotas to other places, it is a sign that these communities do not have any comparative advantage in fish processing, for otherwise the quota owners would find it profitable to land their fish there. Furthermore, the present-day lifestyle in rich Western countries typically involves both spouses working outside the home. Small, isolated communities with few employment opportunities are not magnets for young people looking for a place to establish a family, and they seldom provide the variety of services to which modern men and women feel entitled. Such communities are likely to have found themselves leaning against the wind anyway, irrespective of how the fishing industry is organized. That notwithstanding, people living in such communities and their representatives will use their political influence to battle against private use rights in fisheries if they see them as a threat. Many of the opinions sought by the Norwegian Ministry of Fisheries in connection with its plan to introduce ITQs came from such communities. Bureaucrats serving the fishing industry in various ways might use their influence for a similar purpose. They may thrive better on the industry's troubles than on its success, particularly if the latter comes through self-regulation diminishing the need for public regulation.

But even those who would receive use rights sometimes oppose them. A part of the explanation lies in controversies over how rights are to be distributed. It will typically be difficult to do so in a way that satisfies everybody. Criteria based on catch history disfavor those who have recently entered the industry, and giving the same amount to all ignores those who work harder and are better skilled than others. The formulas used for allocating fish quotas among those found eligible have been of various kinds, but most have tried to balance the "historical" approach and the interests of newcomers with a significant investment. And eligibility itself, not to forget, has often been difficult to resolve. Laws and regulations seldom take immediate effect once they are signed. More important, they can take a long time to prepare. Firms and individuals who see new laws and regulations take shape will make the necessary adjustments and dispositions to maximize both the probability of being eligible and the share of the pie they are going to get. This positioning game can be costly. Imagine what would have happened in the Alaska halibut fishery, with fishing seasons already down to two frenzied days, if the cutoff date for eligibility had been left open. In fact, some fishermen are rumored to have entered the fishery after 1990 precisely because they did not believe that the deadline would be maintained, given how long it would take to put the program into effect. In the surf clam fishery on the Atlantic coast of the United States vessels were reactivated and fishing effort increased in anticipation of quotas handed out on the basis of catch history, an anticipation that turned out to be largely correct.[1]

The competition for the largest possible share of the gains from rights-based fishing may in fact take on a higher priority than realizing these gains in the aggregate. In other words, fights over the design of a rights-based management system could get in the way of putting the system in place. This is particularly likely to happen if the industry is highly differentiated (small boats versus large, different types of gear) and characterized by mistrust and discord. If some part of the industry thinks it enjoys greater political sympathy than other parts, it is likely to try to gain through politics what it would not be able to gain through the marketplace. The small-boat operators in Iceland have played this card with much success over the years. This part of the industry will typically be against market-driven arrangements in which survival is determined by profitability instead of by political wheeling and dealing in committees, councils, or government offices.

One argument that would seem to favor fishing rights, and ITQs in particular, is that fishing would become less uncertain. Fishing is a highly risky business. It is probably the industry with the highest fatality risk in the world, and success depends critically on forces of nature over which humans have no control, such as the weather and the abundance and migrations of fish, all of which are or can be highly variable from year to year. Having a fixed share of an overall catch quota eliminates much of the uncertainty about how much one will catch, except for the weather (will it be possible to catch the fish one is entitled to?) and whether fish stocks have been assessed with reasonable accuracy (will there be enough fish available to take the quota?). Having a fixed quota share also eliminates the incentive to venture to sea in bad weather to catch the fish before somebody else does. This argument has loomed large in the debates on the merits of ITQs in the United States, but the evidence as to what happened after ITQs were put in place is somewhat mixed.[2] But in some people's view this reduced uncertainty is an argument against rather than for fishing rights. There are those who think that luck will be on their side so that they would come out better in a derby-style competition. Some fishermen are on record for resenting the chance element being removed from the fishery. And there are those who realize that bad luck may put them out of business, in which case they might not be well served if the price of reentering the industry has gone up.

Political Obstacles to Use Rights

Establishing private use rights in fisheries is a political process requiring legislative and regulatory action by the state. This process may run into difficult obstacles and may not even get started. At first glance this may appear surprising. The social gain from privatized use rights lies in greater efficiency; capital and labor otherwise wasted on redundant fishing boats would be directed to more productive uses, and the value of the product itself is likely to be enhanced. (The latter has been a prominent effect of some use-rights programs in the United States.) So, in order to be interested in establishing use rights in fisheries, governments and legislative assemblies would have to be interested in such efficiency gains. Why would they not be?

There are several reasons why governments might not be particularly interested in enhanced efficiency in the fishing industry, or in other industries for that matter. In most countries the fishing industry is such a small part of the economy that it hardly matters one way or the other. The effect on the gross domestic product of eliminating waste in the fishing industry would hardly be visible. The United States is a good example; the fisheries are important locally but not overall; even in the state of Washington, home to much of the Alaska pollock fleet, they do not amount to much in comparison to giant corporations such as Microsoft and Boeing. In the United States there is, on the other hand, an increasingly vocal lobby of environmentalists who see the fishing industry as an unwelcome intruder, particularly those parts of it that capture fish with large and efficient vessels. The influence of this lobby is noticeable in several ways. The emphasis on economic efficiency was considerably watered down in the latest revision of the Fisheries Management Act (now known as the Magnuson-Stevens Act). The activities of the industry have been increasingly constrained for the alleged benefit of sea lions and other animals. Plans are afoot for closing large areas of the sea to fishing altogether, to protect fish as wildlife. The establishment of marine protected areas is a part of an ongoing process in the United States, driven by environmentalists, to set aside large areas in which wilderness is protected and extraction of natural resources is not permitted.

By contrast, governments in countries or jurisdictions where the fishing industry is an important part of the national economy should be the ones most interested in establishing exclusive use rights in their fisheries. There are not many countries like that. Iceland is one of them, and practically all fisheries within the Icelandic 200-mile zone are regulated by ITQs. So far at least, a critical mass of Icelandic politicians and members of the public apparently realize that making the fishing industry an employer of last resort would be inimical to economic growth and welfare. New Zealand may fit this picture as well; the fishing industry is not very large as a part of the total economy, but it is an important export industry. If this can be generalized, the conclusion would be that the closer the political level where the necessary legislative and regulatory action takes place is to those whose incomes and welfare depend on the fishing industry, the more likely it is that rights-based, efficiency-enhancing systems will be established. The more diluted the benefits of any legislative or regulatory action, the less likely it is to occur. But if this is the rule, it is not without exceptions. It has been argued that ITQs were established in the rockfish fishery on the Pacific coast of Canada precisely because it was sufficiently unimportant at both the federal and the provincial level.

Given the emphasis the coastal states of the world put on establishing the 200-mile zone as a part of international law, it is a bit surprising how little interest many of them seem to take in promoting economically efficient utilization of the fish resources inside their zone. The Canadian government, prominent in the battle for coastal states' rights, has no explicit policy on promoting ITQs or other use rights in Canada's fisheries. Apparently, many governments perceive other benefits than economic efficiency from their fisheries. Some governments see fisheries as a generator of jobs in economically depressed areas. From a wider perspective this is wasteful; it would be better to encourage people to move to areas where their labor is in demand, or at the very least not to discourage them from doing so. In the worst of cases, preserving jobs in the fishing industry has proved to be utterly self-defeating. The Canadian government poured money into keeping fishermen in Newfoundland until the cod disappeared. That, sad to say, may be the best structural help the Newfoundland economy got for a long time; people have now started to leave this economically depressed province for better opportunities in other Canadian provinces. The European Union has diverted large amounts of money to its fisheries, thereby promoting overcapacity of fishing fleets and depletion of fish stocks. The main reason has been the EU's political need to transfer money to its poorer members, particularly Spain and Portugal. The aim may be praiseworthy, but the consequences have been dismal, although not as disastrous as they ultimately turned out to be in Newfoundland.

This perception of the role of the fishing industry as an employer of last resort in depressed areas rather than a contributor to the national wealth is one example of how politicians may choose to promote regional, sectoral, and other special interests rather than the interest of the general public. This is not entirely surprising. The careers of politicians in representative democracies depend on their being elected and reelected. It is less obvious than one might think that their chances of being elected depend on how well they have promoted the public interest. On the contrary, their chances might depend more critically on their ability to serve special interests, which often go against the public interest. There are two reasons for this. One is the asymmetric distribution of general and specific interests, or consumer and producer interests. All of us have a common interest in efficient production and cheap imports to keep down the prices of the things we buy. But our ability to pay depends on our income, and our income is tied to a specific, producer interest. A farmer will welcome low prices of things he buys but he is more interested in high prices for his products, which in some countries are achieved through subsidies and import restrictions. Individuals seeking to be elected to represent rural areas will do well, therefore, to promote such subsidies or import restrictions. A steelworker would appreciate lower food prices resulting from cutting agricultural subsidies or allowing free imports, but he is likely to be much more concerned about the viability of the steel mill where he works. Therefore, if he has a choice between a representative who argues for free trade and abolishing subsidies and one who argues for protection against steel imports and other imports, he will probably support the latter.

The other reason why elected representatives have incentives to support specific interests rather than general interest has to do with the financing of political campaigns. The electorate must be mobilized; the incentive for each voter to cast his or her vote is very weak, the vote being one among many and therefore almost without any influence at all, so it is perhaps a surprise that anyone bothers to vote. In some countries voting is actually compulsory. More important, perhaps, the candidates must tailor their message to their audience. Their chances of getting elected depend in no small way on how well they misrepresent their intended policies and how well they tailor their presentations to audiences with different interests. Campaigning for office requires funds, and the funds must be solicited among those who have them, in particular major corporations, industry associations, and large trade unions. The donors will, needless to say, expect something in return, and will hardly support those whose policies run counter to their interests. If a company's interest and the public interest coincide, this would, of course, be fine, but such is not necessarily the case; companies have an interest in monopolizing markets, farmers' or fishermen's associations in procuring subsidies and restricting imports that compete with them, and labor unions in preventing changes that might destroy their members' jobs. An astute politician will have to calculate carefully how to balance the interests of his donors and the interests of the voters.

Because sectoral interests often prevail over the public interest, it is critical for establishing use rights in fisheries that the industry perceive such rights as being in its interest. If the support for such institutional innovation is not coming from the industry, it is indeed hard to see where it would otherwise be coming from. The average voter probably does not much care, as he or she does not have much at stake. To the extent that the general public cares, it is likely to be concerned with other things than the wealth-generating ability of the industry—for example, American environmentalists worry about the sea lions and the dolphins. Not surprisingly, no system of use rights seems to have been put in place without a critical support from the industry, and proposals without such support have foundered. In some cases of success, the industry appears in fact to have been the major driving force.

Distributional Obstacles

The consequences of open access to a limited resource have long been well understood. Over the last quarter century or so, a number of fisheries have been opened up, fishing previously unexploited stocks. If there is merit in all the studies of open-access regimes, they should have had a bearing on how these new fisheries were organized. In some cases they did. The New Zealanders introduced ITQs in their newly developed deepwater fisheries at an early stage. Among those who advised them was Lee Anderson, a renowned American fisheries economist who later played an important role in getting ITQs going in the surf clam and ocean quahog fishery. The wreckfish fishery in the United States was quickly put under ITQs, partly through the efforts of economists working for the fisheries management authorities involved. In other cases, opportunities have been missed. This happened in Chile and in the Alaska pollock fishery, and in Iceland it took almost 10 years to get a rights-based regime in place after the foreign trawlers were expelled and the Icelanders got full control of the demersal stocks around the island. One could have hoped for a quicker learning, or a more expeditious use of received wisdom.

More typically, ITQs have been put in place after a fishery has reached a crisis and other regulations have proven inadequate. Even then the disputes over initial allocations and other design features of the proposed system have gone on for years. In the meantime the situation has gotten worse. This has happened in the United States, in Iceland, and in Chile.

But sometimes the success of ITQs could be their undoing. The demand for crewmembers falls as redundant fishing boats are withdrawn from the fishery, some people may be thrown out of work, and labor unions may lose some of their bargaining strength. Less employment in oversized fisheries is not only natural but desirable; after all, excessive fishing effort means unproductive use of labor that should be directed to other purposes. We may deplore boatowners getting richer at the expense of crewmembers, but the latter are not necessarily underpaid relative to their skills. If crewmembers anticipate this they will use their political clout to prevent ITQs or other exclusive use rights from being put in place; if this is the experienced outcome, they will seek to have such measure removed.

Similarly, fish processors might lose if overcapacity in the fish processing sector has been built up to deal with peaks in landings under competition for a given catch. Getting rid of this excess processing capacity is a natural and desirable consequence of moving to a more efficient management system and hardly an argument for preserving the old and wasteful ways. But in some fisheries, in the United States in particular, expected or actual losses by fish processors have turned out to be major obstacles against ITQs and have sometimes been used as explicit arguments against such systems.

Another pitfall that success may create is the rent generated by a successful system of use rights. The rent will become capitalized in a market value of quotas or boat licenses and will, roughly speaking, be cashed in by those who initially got these rights, to the extent that they did not have to pay their full value (and in most cases such rights have been handed out for free). This is not a gain taken at anybody's expense (except that crews might become worse off) but one generated by a better management system. Letting this potential gain be absorbed by waste of manpower and capital, as under open access, is hardly the best way to deal with its distribution. There are better methods available, such as rent taxes and auctions of quotas, which preserve the efficiency gains but distribute them more widely. Those who gain from a quota system are, however, in a situation similar to the settler who comes to an unsettled land. He who first takes the land into possession will augment his wealth correspondingly, a process that those who come after the land is settled cannot expect to repeat. A rent capture system that is put in place after the initial rights holders have disappeared or sold out could amount to adding insult to injury, instead of a step toward greater fairness.

But one might never get as far as allowing success to generate its own undoing. Inefficiencies such as develop in open-access fisheries generate their own constituencies for keeping things as they are and preventing any efficiency gains from being attained. In principle the gains to be realized should suffice to compensate all potential losers, but in practice full compensation of every loser is seldom possible. Those who are left out, or think they will be left out, or think they should have been given more, will mobilize for opposition. Sometimes they succeed. In any case, the process leading to ITQs in Iceland and the United States has not only been delayed by these forces; to an extent, it has been shaped by them. And in some countries the process has ground to a halt. This underlines the desirability of setting up an ITQ regime, or whatever rights-based regime is appropriate, as quickly as possible, particularly in a "greenfield" situation, before recalcitrant interest groups with a stake in inefficiency have emerged. One notable neoliberal reformer, Sir Roger Douglas, formerly minister of finance in New Zealand, put it this way:

Do not try to advance a step at a time. Define your objectives clearly and move towards them in quantum leaps. Otherwise the interest groups will have time to mobilize and drag you down. Once the program begins to be implemented, don't stop until you have completed it. The fire of opponents is much less accurate if they have to shoot at a rapidly moving target. Consensus among interest groups on quality decisions rarely, if ever, arises before they are made and implemented. It develops after they are taken, as the decisions deliver satisfactory results to the public.[3]

Should Losers Be Compensated?

Structural changes that impose costs on some individuals who in the absence of compensation would be worse off than if they had been able to continue operating in the old way are not confined to the fishing industry. Computers and direct dialing have made hordes of secretaries and switchboard operators redundant and their skills obsolete. The automobile drastically reduced the demand for makers of wagons and saddles; the steamship did the same to sail makers and rope makers. That notwith-standing, few of us would probably want to return to these old ways, even if they may have had their charm. Anybody who has seen a sailing ship head toward the horizon will have to admit that, esthetically, the steel chunks now traversing the oceans are far inferior. Nevertheless, there is little doubt that the said changes produced a huge surplus of benefits, even if certain occupational groups are likely to have been made worse off than if they had never happened.

Given that there is a net benefit from changes like these, and from better fisheries management for that matter, the idea is not far fetched that a part of this surplus should be used to compensate those who lose from the change. Some would even go as far as saying that no change is justified unless all losers are compensated, ensuring a fair distribution of the benefits.

Institutions such as the American Fisheries Management Councils would seem well suited to ensure that this indeed takes place. The process is trans-parent and open, and those who have opposed changes such as ITQs have taken advantage of this and aired their arguments with much vigor and held up the process for years. But precisely this makes one wonder how desirable or practicable full compensation of alleged losers is. Some of the opponents of the Alaska halibut quotas did not seem to have a lot to lose, and sometimes their alleged losses were questionable. Some had entered the fishery after the so-called qualifying period in full knowledge that they were not eligible for quotas, counting on a perceived political need to extend that period because of many new entrants. Some had declined to take part in short and disadvantageous openings. And some had little at stake but were generous with their advice.

These examples make us look back and wonder how many of the tech-nological or institutional innovations that have brought us the benefits of modern life would have been put in practice if they had been hostage to a process committed to "transparency and stakeholders' representation" (expressions that are much in vogue, and not just in circles dealing with fisheries policy). The wagon makers' and saddle makers' unions would probably have found many faults with the automobile. It could easily catch fire, and it was noisy and smelly. There were few filling stations. And what if there were troubles with the engine? With luck they might have dis-missed this creature as too slow and uncomfortable to be much of a threat.

The switchboard operators' union would not have liked direct dialing. Typesetters were made obsolete by the computer and advances in printing technology. In the 1970s and the 1980s, major newspapers were held up repeatedly because of strikes among typesetters trying to protect their jobs. The newspaper magnate Rupert Murdoch dealt with that in his own way; he secretly prepared new printing works in one of the suburbs of London and suddenly, one day, the whole operation was moved there, leaving the typesetters with nothing to do and unable to obstruct printing. Nowadays journalists type their text directly into a computer, and the text is cut and pasted with the click of a mouse to fit the page where it appears.

It seems, therefore, that the otherwise so appealing ideal of ensuring that all losers will be compensated before any change is made could easily stifle progress and lock us into outmoded technologies and institutions. This raises tough question: When is it admissible to look at the sum of expected benefits less costs associated with a change and go on with it when the sum is positive? Are we ready to accept any costs as long as they are less than the benefits? The amount, the nature, and the distribution of the costs would seem to matter.

Any society has to resolve for itself how it deals with questions like these, and different societies do so differently. No less important, societies do so differently at different times. What was acceptable 100 years ago is considered not so good today and may be thought barbaric 100 years from now. The course of history has been one of conquests, of ethnic cleansing, and of inferior modes of production and organization being brutally cast aside. Taking a long view, we have made progress, materially at any rate. We have even reached a point where those methods of old are viewed with abhorrence. Yet we would not be where we are without them. Where our present more civilized ways will take us remains to be seen.

Evolution of Institutions

The development of economic institutions is an evolutionary process. An institutional change arises in response to a perceived need; someone—a government official, a politician, an industry player—has to perceive a gain, not necessarily of a personal nature, to be obtained from an institutional change. A design will be proposed. If it is to succeed, a constituency has to be found to carry it forward. What that constituency is will depend on how societies are governed. In democratic societies the support, or at least the acquiescence, of the electorate is necessary. In the hierarchical societies of old one had to secure the support of the ruling class. Money went a long way to buy such influence, and even in present-day democracies it still does. Policy in democratic societies emerges from the joint forces of popular will and the interests of those who finance the political machines. And popular will is molded by persuasion, which costs money.

Any institutional change will affect different interests differently. Each interest group will attempt to make it serve its own special interest. The outcome of this fight can be difficult to predict; it depends not only on the economic strength and cohesion of the groups involved, but also on how well they succeed in appealing to a wider constituency that perhaps has a peripheral interest but yet some influence. In democratic societies this is particularly important. The final "proposal" may be much affected by this fight among interest groups and the need to appeal to a wider audience; it could in fact be quite different from the original idea and perhaps not very well suited to solve the problem it was originally meant to solve. This seemingly unpredictable outcome is analogous to random mutations in nature. Pushing the analogy with nature a bit further, those institutions that survive presumably are such as turn out to serve a purpose. It could even be a purpose different from the one the institution was meant to serve.[4]

Thus, the design of economic institutions is the outcome of bargaining among special interests rather than of adopting an idealized blueprint. Much depends, however, on the relative strength of governments and special interests, and on whether the incentive structure embedded in the economic system is such as to bring special interests into harmony with the public interest. Governments promote free trade against opposition from those who would be adversely affected by free trade not only because it will enhance the general economic welfare but also because exports will benefit certain sectors of the economy and free trade is a two-way street; your exports must be paid for by somebody else's exports, which are your imports. A fishing or a farming industry that has gotten used to subsidies or protection against competing imports will seek more of the same, but if those avenues are effectively closed the industry players will seek to raise their profits by increasing efficiency and by asking for government regulations that improve their ability to do so.

Having economic institutions evolve through bargaining with and between special interests need not be singularly bad, however. Anchoring economic institutions firmly among practical men of business provides a certain guarantee that they will actually work. How well they will serve the public interest depends very much on how well private and public interests are aligned, on how the economy is organized, and on whether the state is predatory, a source of favors, or a keeper of the rules of the game and a provider of public services (without which no market economy can work). Economists and other academics with ideas that look nice enough on paper but are unworkable have never been in short supply. A few generations ago the phrase "national economic planning" was common currency with positive connotations among economists. That phrase is seldom heard nowadays. No idea has fallen so totally into oblivion as the notion of having government agencies plan and run whole economies.

The development of exclusive use rights to fish is an interesting example of the evolution of economic institutions. The need to establish use rights in fisheries arose from the increased scarcity of fish, which was due to technological progress in fisheries. Countries wanted to secure for themselves the wealth to be created from fishing, and lately industry players have started to play the same game against new entrants into "their" fishery. Theory and experience from other arenas provided blueprints for the design of exclusive use rights; however, since they were to be applied in a new and in some ways unique setting, there was much uncertainty as to what would work and how. These rights also had to be embedded within the existing system of rights and traditions. At this point, experiment and evolution take over; the political process, a battle of different interests whose outcome is far from predetermined, molds the design. The 200-mile limit owes much to a twisted and torn conception of a neutrality zone having nothing whatsoever to do with fishing and fish stocks, but it serves the national interests of a number of states. A quite different blueprint existed (the Truman Proclamations), but one serving a not dissimilar purpose. Different exigencies, by no means all related to fishing, ensured the codification of the 200-mile limit.

On the basis of the 200-mile limit, variants of exclusive use rights in fisheries have been developed. The theory and exploitation regimes for other natural resources provided guidance for a blueprint. Production quotas are not uncommon in agriculture and were a model for the Dutch fish quota system, initiated in the 1970s.[5] The stinting of the English commons inspired the Canadian economist Peter Pearse, an early proponent of ITQs. But the blueprints have been variously modified in response to different interest groups and popular perceptions. These modifications can be seen as "mutations" whose success will have to be proven by their practicality.

As in the evolutionary processes in nature, these new institutions will survive if they turn out to serve a purpose. What that purpose will be is not entirely given and may also evolve over time. The fishing industry will support regimes based on use rights if they are in its interests—that is, if they generate rents captured by the industry. There is little doubt that they do; after use-rights systems have become established, they have resulted in a rising value of the assets to which these rights have been attached, be they fish quotas or vessel licenses. This has made these systems more entrenched; the owners of these assets would not happily see their value eroded. But the industry is embedded in the society of which it is a part. If society is to support use-rights systems, it has to perceive a benefit therefrom or a purpose therein. In countries where the fishing industry is an important source of wealth, use rights are likely to be supported because they promote wealth generation. But this is not a foregone conclusion. Some people see fisheries as a common resource that should be open for all, in the name of fairness and equity, and are oblivious to the fact that this may be the surest to way to common ruin. The dire experiments with socialism in the twentieth century arose from such misconceptions.

In societies where fisheries are so unimportant as not to be noticed, the industry is likely to be able to call the shots, with public indifference. But neither is this a foregone conclusion. An increasing part of the public, otherwise remote from anything having to do with fish and fisheries, is taking a growing interest in fish as wildlife and not as a source of material wealth. Exclusive use rights have no role to play in that universe; fishing is an activity that preferably should be gotten out of the way but may tolerated if it is pursued by primitive methods and some cute ethnic minorities. The greatest threat to fisheries in the future may not be overfishing and depletion of fish stocks but rampant environmentalism.

Notes

Introduction

1. The great auk, a sea bird that could not fly very well, was hunted to extinction in Iceland in the nineteenth century. The buffalo roamed over the American prairies in large numbers as late as the nineteenth century, but was hunted indiscriminately, and now there are only few left. Buffalo Bill alone killed more than 4,000 of them in 8 months for feeding construction workers on the Union Pacific Railroad.

Chapter 1

1. The quotations are from pp. 828–832 of the fifth edition of Paul A. Samuelson's *Economics* (McGraw-Hill, 1961).

2. Rousseau 1755, p. 180. I am indebted to Hannes H. Gissurarson of the University of Iceland for this quotation and those that follow.

3. Ibid.

4. Ibid., pp. 45–46.

5. De Soto 2000.

6. Quoted from p. 137 of Hirshleifer 1982.

7. Schelling 1978.

8. In 2003 the Swiss parliamentarian Ruth-Gaby Vermot-Mangold delivered a report on such trade in organs to the Council of Europe, after a visit to Moldova. Moldova is by no means the only country providing such supplies.

9. Being the king's son was also risky. Fratricide was not uncommon in royal dynasties, particularly when the succession did not necessarily go to the eldest son.

10. These rules are recorded in the *Landnámabók* (Book of Settlers), quoted here from p. 45 of Jóhannesson 1956.

11. De Soto 2000.

12. See, for example, Klebnikov 2000; Freeland 2000.

13. An authoritative work on the English enclosures is Gonner 1912.

14. Quoted on p. 309 of Gonner 1912.

15. Hardin 1968. This insight is old, however. Aristotle wrote: ". . . that which is common to the greatest number has the least care bestowed upon it. Everyone thinks chiefly of his own, hardly at all of the common interest; and only when he is himself concerned as an individual." (*Aristotle's Politics and Poetics*, Viking 1974, *Politics*, book 2, chapter 3, p. 27.) David Hume mused over peasants' incentives to dig a dike draining a common pasture.

16. Gonner 1912, p. 103.

17. Thomas Becon, *Jewels of Joy*, circa 1540, quoted from p. 82 of Scrutton 1887.

18. Quote from 1622. See Scrutton 1887, p. 104.

19. Gonner 1912, pp. 132–136.

20. Quoted from p. 177 of Gonner 1912.

21. Scrutton 1887, p. 106.

22. Gonner 1912, p. 182.

23. Ibid., pp. 72–73.

24. Ibid., p. 337.

25. Quoted on p. 360 of Gonner 1912.

26. Lord Ellenborough in the House of Lords 1836, quoted from p. 156 of Scrutton 1887.

27. Fagan 2000, pp. 106–110.

28. Gonner 1912, p. 299. This was not without foundation. As late as World War II, the hedgerows of Normandy constituted a major obstacle for the advancing Allied forces.

29. Scrutton 1887, p. 155.

30. On the history of the Highland clearances, see Richards 1982, 1999.

31. Richards 1982, p. 50.

32. Richards 1999, p. 293; Richards 1982, p. 306.

33. Richards 1999, p. 271.

34. Ibid., p. 285.

35. Ibid., p. 258.

36. Ibid., p. 40.

37. Ibid., p. 43.

38. Richards 1982, p. 182.

39. That is probably in no small measure due to the moderating influence of a Scottish gentleman and a member of Parliament, James Loch, who managed Lord Stafford's business interests. Those who dislike the feminist-inspired replacement of the traditional "fisherman" by "fisher," which over the last 20 years or so has entered the vocabulary of political correctness, may take comfort in the fact that Loch used "fisher" in a letter he wrote in 1818 (see Richards 1982, p. 319).

40. Richards 1982, chapter 15, especially pp. 495–496.

41. Gonner 1912, p. vii.

42. Demsetz 1967.

43. This debate has been eloquently summarized by Tim Smith (1994).

44. Fulton 1911.

45. Ruddle 1984.

46. Ruddle and Johannes 1984.

47. Fulton 1911.

48. Fulton 1911.

49. See, for example, the diagram of catches of Icelandic cod in figure 7.1.

Chapter 2

1. Quoted on p. 495 of Hollick 1977.

2. Fulton 1911, p. 520.

3. *Mare Liberum, sive de Jure quod Batavis competit ad Indicana Commercia Dissertatio.* Quoted from p. 342 of Fulton 1911.

4. Fulton 1911, p. 563.

5. For this story, see Armanet 1984. Armanet presents the Panama Declaration as an agreement between the United States and the UK. This is not correct; the declaration was made at a meeting between the foreign ministers of the American

republics (see *Department of State Bulletin* 1, 1939: 331–333). It is unlikely that there was any understanding between the US and the UK about this; in December 1939 the UK defied the neutrality zone and chased a small German warship, the *Admiral Graf Spee*, in the zone until it took refuge in Montevideo. The story is also told in the article *Las Doscientas Millas* by T. Barros in the Chilean daily *El Mercurio de Santiago*, July 19, 1974, which refers to the map in Semana Internacional. For more on the origins of the 200-mile limit, see Hollick 1977, which also refers to the map in Semana Internacional, possibly based on the article by Barros.

6. The map is published in *Foreign Relations of the United States: Diplomatic Papers*, Department of State, United States of America, volume 5 of 1939 series.

7. Hollick 1977, p. 498.

8. Bailey 1980, p. 687. The first edition of this book was published in early 1940.

9. Fulton 1911, p. 575.

10. The euphemism "developing countries" has become so entrenched that it will be used here occasionally. It refers, as most people know, to countries with a low GDP per capita, but, sadly enough, many of them have stagnated or even moved backward in terms of economic development.

11. Eckert (1979, pp. 278–279) mentions an entire morning wasted at Caracas on praising Simon Bolivar, a suspension of proceedings in Geneva in March 1975 for paying tribute to the late King Faisal, and in New York in September 1976 because of Chairman Mao's death. He also laments the verbiage; at the Caracas session 90 mimeograph operators worked around the clock on 27 machines to produce 250,000 pages of daily documentation for the delegates.

12. On the second US proposal, see p. 36 of Burke 1994.

13. Pohl 1985, p. 39.

14. A standard reference on the operations of the Third UN Law of the Sea Conference is Nordquist 1985–. Another informative text is Miles 1998.

15. There are persistent rumors to the effect that there was never much fishing going on in the Donut Hole, but that it provided a cover for foraging into the US and Russian zones, as these areas cannot be effectively controlled. Better controls (satellite tracking) and political pressure rather than depletion of fish stocks could have prompted the countries fishing in the Donut Hole to enter into agreement with the US and Russia. The "cover" provided by holes and areas contiguous to the economic zone may indeed be a more important argument for a further extension of the zone than the fishing that takes place in these areas.

16. Plé 2000; Burke 2000.

Chapter 3

1. Some readers will have noticed that this amounts to assuming that all boats are identical. Allowing for difference among boats does not greatly affect the conclusions about to be drawn.

2. This can be seen as follows. The total sustainable catch (C) can be written as $f(B)B$ where $f(B)$ is the sustainable catch per boat and B is the number of boats. The sustainable contribution of an additional boat to the total catch is the derivative of C with respect to the number of boats: $dC/dB = f(B) + (df/dB)B$. Since $df/dB < 0$ (the sustainable catch per boat falls as more boats are added), $dC/dB < f(B)$ and could be negative.

3. Recreational fisheries are a different thing; the benefits of those fisheries are largely or wholly immaterial and do not fit within the framework of analysis in this book, which is concerned with commercial fisheries yielding material benefits.

4. This discussion ignores the effect of discounting, which justifies sacrificing some sustainable yield for an unsustainable short-term gain. It can be shown, however, that the open-access solution would not be optimal unless the discount rate were infinite, so no great harm is done by focusing on sustainable yield.

5. Rent is the difference between the revenues from fishing and all necessary costs to obtain the fish. The concept of rent applies to scarce natural resources; land, minerals, even talent (the clever lawyer or the exceptional singer can sell their services for a higher price than less skilled competitors).

6. To see this, note that when the rent is maximized, the marginal revenue is equal to the marginal cost. The marginal revenue is what we have termed here the contribution of an additional boat while the marginal cost is constant and equal to the cost per boat.

7. Compare Garibaldi and Limongelli 2002, p. iv.

8. Weitzman 2002.

9. These points are further elaborated in Hannesson and Kennedy 2003.

10. This is not true under all circumstances; it can be shown that when the crew is remunerated by a share of the catch value and not with a fixed wage there will most likely be an incentive to overinvest (Hannesson 2000). Even so, it is likely, to say the least, that overinvestment will be much less of a problem with ITQs than without them.

11. Robinson 1986.

12. Iyambo 2000.

13. Gislason 2000.

14. Vetemaa et al. 2001.

15. Figure 3.3 shows the total catches of anchoveta (Peruvian anchovy) and illustrates the short-term fluctuations we are dealing with here. The anchoveta also occurs in Chilean waters and is caught by Chilean boats.

16. Such cases are the Atlanto-Scandian herring (Ulltang 1980), the South African pilchard (Butterworth 1981), and the Northern cod of Newfoundland.

17. Valatin 2000.

18. Stollery 1986.

19. For example, open fishing boats in Iceland are equipped with several computer-controlled jigging machines with several hooks attached to each. The machines jig the hook lines and pull them up automatically when enough fish have gotten hooked. When the fishing is good, the fisherman on board may have his hands full in getting the fish off the hooks, putting them in storage, and throwing the lines out again.

20. A good example of this is the technological leap in the Northeast Atlantic herring fisheries in the 1960s. The so-called power block made it possible to pull purse seines mechanically instead of by hand. This in turn made it possible to use much bigger seines and boats. The introduction of this technology occurred over a few years in the early 1960s and increased the capacity of the fishing fleets many times over. The catches of herring increased rapidly as a result, and the herring stocks crashed around 1970, most likely because of overexploitation.

21. As an example, a regulation was introduced in Norway permitting small trawlers to fish within the old 12-mile territorial limit. At the time the regulation was issued the so-called large trawlers were well above the size limit, which was 300 GRT (gross registered tons). As the fleet was renewed, the new vessels typically measured 299.9 GRT.

22. See, for example, "Fiskaren," November 1, 2002, p. 2.

23. There exists considerable literature on TURFs. One important contributor is Francis Christy (1983, 2000). The conference proceedings (Shotton 2000a,b) contain several papers dealing with TURFs, with extensive references.

24. The most comprehensive effort in this regard is Baland and Platteau 1996.

25. Christy 2000, p. 126.

26. Asada, Hirasawa, and Nagasaki 1983; Japanese contributions in Yamamoto and Short 1992.

27. Christy 2000.

28. In many parts of the world this is still a reality. Attacks from crocodiles are among the most common causes of fatalities among Tanzanian fishermen, and workers on rubber plantations in Malaysia are occasionally attacked by tigers.

Chapter 4

1. A fourfold increase implies an annual growth rate of 15 percent. Hence, an inflation rate of 5 percent implies that one-third of the growth would be due to inflation.

2. Christy 1973.

3. The papers were published in a special issue of the *Journal of the Fisheries Research Board of Canada* in 1979. That journal is now known as the *Canadian Journal of Fisheries and Aquatic Sciences*.

4. Moloney and Pearse 1979.

5. For the proceedings of this conference, see Sturgess and Meany 1982.

6. The European Union is an interesting exception. Its member states, though sovereign, do not have exclusive rights to fish within the 200-mile zone; those rights belong to the Union. Interestingly, rights to the resources of the continental shelf belong to the individual states. This is but one example of how the European Union is a halfway-confederation. Its member states have ceded to it some of their sovereignty, although usually not that which is most vital to their national interest and most prestigious (for example, foreign policy, offshore petroleum). An exception is the euro and the European Central Bank, but some of the member countries, most notably the United Kingdom, still retain their own currency and central bank.

7. On this, and on Canadian quota regimes in general, see Burke and Brander 2000.

8. Acheson 1979.

9. Ruddle and Johannes 1984.

10. Robinson 1986.

11. Steen 1930, p. 30.

12. Since 1999 it has been possible to use ITQs as lien in Iceland.

13. De Soto 2000.

14. Tsamenyi and McIlgorm 2000.

15. *Federal Reporter*, third series, volume 161, p. 588.

16. *Federal Supplement* 762: 375–376. For more on the legal discussion, see Nielander and Sullivan 2000.

17. Ostrom 1990.

18. Stevenson 1991.

19. These examples are from Christy 2000.

Chapter 5

1. See Bess 2000 and references therein.

2. The Maori are the indigenous people of New Zealand. They are of Polynesian origin. Their ancestors are believed to have come to New Zealand about a thousand years before the British settled in the country.

3. Hannesson 1989.

4. Hersoug 2002. Hersoug also describes these modifications in some detail and provides an interesting analysis and evaluation of the New Zealand management system.

5. Newell, Sanchirico, and Kerr 2003, p. 9.

6. Batstone and Sharp 1999, p. 186.

7. Kidd 2000, p. 141.

8. On industry and management costs, see Schrank et al. 2003.

9. Batstone and Sharp 1999, p. 183.

10. Batstone and Sharp 1999, table 5, p. 184.

11. Newell, Sanchirico, and Kerr 2003.

12. Much of the following discussion on Chile is based on work by Julio Peña-Torres (1995, 1997, 2002).

13. After a recent merger, the Angelini group now controls 80–90 percent of this fishery, which makes it an interesting object for study of the effects of sole ownership.

14. ITQs were, however, not the sole reason for crew layoffs; some layoffs had already occurred before introduction of ITQs during the strong El Niño of 1997–1999.

15. This was the conclusion of a study of the pelagic fisheries (T. Hansen et al., En strukturanalyse av sildolje- og sildemelindustrien, Center for Applied Research, Norwegian School of Economics and Business Administration, 1976).

16. Cargo capacity is a crude measure of the capacity of a fishing fleet, but it is undoubtedly an important component thereof. The actual capacity of the fleet is

likely to have increased since 1989 through the installation of new and better equipment of various kinds, but no detailed analysis has been done on that issue.

17. An example: An owner of an 8,000-hectoliter boat who bought a 5,000-hl boat would have increased his base quota from 410 tonnes to 480, or by 70 tonnes, according to the quota allocation rule. According to the rules now in force he would keep at least 60 percent of the base quota of the 5,000-hl boat, which is 204 tonnes, for 13 or 18 years. The share the boatowner would keep depends on whether the boat is sold from the northern to the southern part of Norway or vice versa, or within the northern or the southern district.

18. Mikalsen and Jentoft 2003.

19. The main sources of information on these fisheries are Turris 2000 and Rice 2003.

20. Turris 2000.

21. Rice 2003, p. 1.

22. Rice 2003.

23. According to Rice (2003), the work force in the fishery was reduced by one-third as a result of the ITQ regime.

24. Some of the information on the B.C. halibut fishery is from Porter 1996.

25. Grafton, Squires, and Fox 2000, pp. 686–687.

26. Ibid., p. 689.

27. Ibid., p. 709.

Chapter 6

1. The law does not explicitly claim ownership of resources above the continental shelf; in fact it is titled Lög um vísindalega verndun fiskimida landgrunnsins (Law on scientific protection of the fishing banks on the continental shelf). Implicitly, however, the law asserts national ownership, in that it authorizes the Ministry of Fisheries to determine fish protection zones over the continental shelf where all fishing is subject to Icelandic rules and monitoring. Subsequent extensions of the Icelandic fishing limits were based on this law.

2. Calculated from table 3.1.10 in *State of Marine Stocks in Icelandic Waters 2002/2003*, Hafrannsóknastofnunin (Marine Research Institute), Reykjavík, Fjölrit No. 97, 2003.

3. According to the factor $e^{-(F + M)}$, where F is instantaneous fishing mortality and M is natural mortality. The value $M = 0.2$ is frequently used but not well known and

undoubtedly variable over time and between age groups. Likewise the different age groups are hit differently by the "average" F. Nonetheless, this seems an unduly high rate of exploitation.

4. Útvegur 2002, p. 23 (Statistics Iceland).

5. This review is included in an interim report from the "Natural Resource Committee" appointed in 1998 (*Audlindanefnd:* Áfangaskyrsla med fylgiskjölum, pp. 159–236, especially pp. 174–175, The Prime Minister's Office, March 1999).

6. Ibid., pp. 174–175.

7. These issues are discussed in greater detail in chapter 3.

8. Author's own translation of parts of part IV of the verdict, published in the Reykjavík daily *Morgunbladid*, December 4, 1998, p. 12.

9. See, in particular, an article by Jón Steinar Gunnlaugsson in *Morgunbladid*, December 19, 1998, p. 61.

10. With the exception of small boats managed by a limit on fishing days.

11. This was acknowledged by Jón Steinar Gunnlaugsson in his article. The Icelandic words used are "veidileyfi" (Paragraph Five), which can be translated as a fishing permit, and "veidiheimild" (Paragraph Seven), which translates into authorization to fish. The Ministry of Fisheries had been using these words specifically so that "veidileyfi" meant a general permit to fish, whether or not for fish under quota, while "veidiheimild" meant a right to catch a certain quantity of a specific type of fish under quota, determined by the vessel's share of the total allowable catch.

Chapter 7

1. This act, revised in 1980 and in 1996, became known, successively, as the Magnuson Act and the Magnuson-Stevens Act.

2. In most documents one encounters the name "steller sea lion." This animal was discovered and named by a biologist with the name of Steller, and some other animals are also named after him. After a while the original meaning of the word became lost on its users, the capital S and the apostrophe were dropped, and the last s got absorbed by the sea. This is a small example of linguistic evolution; when a mistake is repeated often enough and by many enough it becomes a rule.

3. The information on these fisheries is mainly from Wang 1995 and McCay and Brandt 2002.

4. The federal government controls the fisheries in the exclusive economic zone up to the old 3-mile territorial limit, whereas the individual states control the nearshore fisheries.

5. Brandt 2003. Information in this paragraph is taken from that source.

6. Berman and Leask 1994.

7. Crutchfield and Zellner 1962. This publication was reissued in 2003 with comments by a number of economists.

8. NRC 1999, p. 306.

9. The arguments against the ITQs are summarized in the preamble to the program rules published on p. 59375 of *Federal Register* 58 (1993).

10. Annual Report, IFQ Fee (Cost Recovery) Program. National Marine Fisheries Service, Alaska Region, February 2002.

11. Smith 2000.

12. *Report to the Fleet*, National Marine Fisheries Service, Alaska Region, February 2002.

13. Ibid.

14. Berman and Leask 1994.

15. *Federal Reporter*, third series, 84: 343ff.

16. Sources on the wreckfish fishery: Gauvin, Ward, and Burgess 1994; Nielander and Sullivan 2000.

17. Huppert 1988, 1991.

18. Matulich, Mittelhammer, and Reberte 1996.

19. The Anti-Reflagging Act is described in some detail on p. 3–132 of NOAA/NMFS 2002.

20. NRC 1999.

21. Sullivan 2000.

22. Alaska pollock is also fished in the Gulf of Alaska (i.e., south of the Alaska Peninsula). This is a much smaller fishery than the one in the Bering Sea and around the Aleutian Islands. All of the Gulf of Alaska catch is processed onshore in Alaska. The American Fisheries Act only pertains to the pollock fishery in the Bering Sea and around the Aleutian Islands.

23. Felthoven 2002, p. 185.

24. Criddle and Macinko 2000, p. 463.

25. Wilen 2002.

26. Information provided by At-sea Processors Association, Seattle.

27. Matulich, Mittelhammer and Reberte 1996.

28. For a good discussion of this, see Halvorsen, Khalil, and Lawarrée 2000. That paper also appears as appendix D of NOAA/NMFS 2002.

29. NOAA/NMFS 2002, p. 4–156, footnote 35.

30. In the Pacific groundfish fishery there is a license-limitation scheme, implying a cost barrier to entry. A factory trawler would have to buy ten coastal trawl licenses to get one midwater trawl permit for Pacific whiting. In the offshore fishery for Alaska pollock the vessels authorized to participate are named explicitly in the American Fisheries Act.

Conclusion

1. Brandt 2003.

2. NRC 1999, p. 36.

3. Quoted from chapter 1 of Hersoug 2002.

4. Some interesting analogies between evolution of economic institutions and evolution in nature are discussed in Zerbe 1982, especially on pp. 131–165.

5. On the ITQs in the Netherlands, see Davidse 2000.

References

Acheson, J. M. 1979. "Variations in Traditional Inshore Fishing Rights in Maine Lobstering Communities." In *North Atlantic Maritime Cultures*, ed. R. Anderson. Mouton.

Ackroyd, P., R. P. Hide, and B. M. H. Sharp. 1990. New Zealand's ITQ System: Prospects for the Evolution of Sole Ownership Corporations. Report to MAAFFish.

Anderson, L. G. 2002. "A Microeconomic Analysis of the Formation and Potential Reorganization of AFA Coops." *Marine Resource Economics* 17: 207–234.

Anderson, R., ed. 1979. *North Atlantic Maritime Cultures*. Mouton.

Armanet, P. 1984. "The Economic Interest Underlying the First Declaration on a Maritime Zone." In *The Exclusive Economic Zone*, ed. F. Vicuña. Westview.

Asada, Y., Y. Hirasawa, and F. Nagasaki. 1983. Fishery Management in Japan. Fisheries technical paper 238, FAO.

Bailey, T. A. 1980. *A Diplomatic History of the American People*, tenth edition. Prentice-Hall.

Baland, J.-M., and J.-P. Platteau. 1996. *Halting Degradation of Natural Resources*. Clarendon.

Batstone, C. J., and B. M. H. Sharp. 1999. "New Zealand's Quota Management System: The First Ten Years." *Marine Policy* 23: 177–190.

Batstone, C. J., and B. M. H. Sharp. 2003. "Minimum Information Management Systems and ITQ Fisheries Management." *Journal of Environmental Economics and Management* 45: 492–504.

Berman, Matthew, and Linda Leask. 1994. "On the Eve of IFQs: Fishing for Alaska's Halibut and Sablefish." *Alaska Review of Social and Economic Conditions* 29, no. 2.

Bess, R. 2000. "Property Rights and their Role in Sustaining New Zealand Seafood Firms' Competitiveness." In Use of Property Rights in Fisheries Management: Workshop Presentations, ed. R. Shotton (fisheries technical paper 404/2, FAO).

Brandt, S. 2003. Evaluating Tradable Property Rights for Natural Resources: The Role of Strategic Entry and Exit. Working paper 2003–9, University of Massachusetts, Amherst.

Burke, D. L., and G. I. Brander. 2000. Canadian Experience with Individual Transferable Quotas. In Use of Property Rights in Fisheries Management: Mini-Course Lectures and Core Conference Presentations, ed. R. Shotton (fisheries technical paper 404/1, FAO).

Burke, W. T. 1994. *The New International Law of Fisheries*. Clarendon.

Burke, W. T. 2000. Compatibility and Precaution in the 1995 Straddling Stock Agreement. In *Law of the Sea*, ed. H. Scheiber. Martinus Nijhoff.

Butterworth, D. S. 1981. "The Value of Catch-Statistics-Based Management Techniques for Heavily Fished Pelagic Stocks with Special Reference to the Recent Decline of the Soutwest African Pilchard Stock." In *Applied Operations Research in Fishing*, ed. K. Haley. Plenum.

Christy, Francis T., Jr. 1973. Fisherman's Quotas: A Tentative Suggestion for Domestic Management. Occasional paper 19, Law of the Sea Institute, University of Rhode Island.

Christy, Francis T., Jr. 1983. Territorial Use Rights in Fisheries: Definitions and Conditions. Fisheries technical paper 227, FAO.

Christy, Francis T., Jr. 2000. "Common Property Rights: An Alternative to ITQs." In Use of Property Rights in Fisheries Management: Mini-Course Lectures and Core Conference Presentations, ed. R. Shotton (fisheries technical paper 404/1, FAO).

Clement, G. 2000. "The Orange Roughy Management Company Limited—A Positive Example of Fish Rights in Action." In Use of Property Rights in Fisheries Management: Workshop Presentations, ed. R. Shotton (fisheries technical paper 404/2, FAO).

Connor, R. 2000. "Trends in Fishing Capacity and Aggregation of Fishing Rights in New Zealand under Individual Transferable Quota." In Use of Property Rights in Fisheries Management: Workshop Presentations, ed. R. Shotton (fisheries technical paper 404/2, FAO).

Cooper, L., and L. Joll. 2000. "The Scalefish Fisheries of Northern Western Australia—The Use of Transferable Effort Allocations in the Management of Multi-Species Scalefish Fisheries." In Use of Property Rights in Fisheries Management: Workshop Presentations, ed. R. Shotton (fisheries technical paper 404/2, FAO).

Criddle, K. R., and S. Macinko. 2000. "A Requiem for the IFQ in US Fisheries?" *Marine Policy* 24: 461–469.

Crutchfield, J. A., and G. Pontecorvo. 1962. *The Pacific Salmon Fisheries: A Study in Irrational Conservation*. Johns Hopkins University Press.

Davidse, W. P. 2000. "The Development towards Co-Management in the Dutch Demersal North Sea Fishery." In Use of Property Rights in Fisheries Management: Mini-Course Lectures and Core Conference Presentations, ed. R. Shotton (FAO fisheries technical paper 404/1).

Demsetz, Harold. 1967. "Toward a Theory of Property Rights." *American Economic Review* 57: 347–359.

De Soto, H. 2000. *The Mystery of Capital*. Basic Books.

Eckert, R. D. 1979. *The Enclosure of Ocean Resources*. Hoover Institution Press.

Fagan, B. 2000. *The Little Ice Age*. Basic Books.

Felthoven, R. G. 2002. "Effects of the American Fisheries Act." *Marine Resource Economics* 17: 101–205.

Freeland, C. 2000. *Sale of the Century*. Little, Brown.

Fulton, T. W. 1911. *The Sovereignty of the Sea*. William Blackwood.

Garibaldi, L., and Limongelli, L. 2002. Trends in Oceanic Captures and Clustering of Large Marine Ecosystems: Two Studies Based on the FAO Capture Database. Fisheries technical paper 435, FAO.

Gauvin, J. R., J. M. Ward, and E. E. Burgess. 1994. "Description and Evaluation of the Wreckfish (*Polyprion Americanus*) Fishery under Individual Transferable Quotas." *Marine Resource Economics* 9: 99–188.

Gislason, G. S. 2000. "Stronger Rights, Higher Fees, Greater Say: Linkages for the Pacific Halibut Fishery in Canada." In Use of Property Rights in Fisheries Management: Workshop Presentations, ed. R. Shotton (fisheries technical paper 404/2, FAO).

Gonner, E. C. K. 1966. *Common Land and Inclosure*, second edition. Keelley.

Goodlad, J. 2000. "Industry Perspective on Rights-Based Management: The Shetland Experience." In Use of Property Rights in Fisheries Management: Mini-Course Lectures and Core Conference Presentations, ed. R. Shotton (FAO fisheries technical paper 404/1).

Grafton, R. Q., D. Squires, and K. J. Fox. 2000. "Private Property and Economic Efficiency: A Study of a Common-Pool Resource." *Journal of Law and Economics* 43: 679–713.

Halvorsen, R., F. Khalil, and J. Lawarrée. 2000. Inshore Sector Catcher Vessel Cooperatives in the Bering Sea/Aleutian Islands Pollock Fisheries. BSAI Discussion paper 2-7-00, Department of Economics, University of Washington.

Hannesson, R. 1989. "Catch Quotas and the Variability of Allowable Catch." In *Rights Based Fishing*, ed. P. Neher et al. Kluwer.

Hannesson, R. 2000. "A Note on ITQs and Optimal Investment." *Journal of Environmental Economics and Management* 40: 181–188.

Hannesson, R., and J. Kennedy. 2003. Landings Fees versus Fish Quotas. Working paper 27/03, Center for Fisheries Economics, Bergen.

Hardin, G. 1968. "The Tragedy of the Commons." *Science* 162: 1243–1248.

Hart, J. A. 1976. The Anglo-Icelandic Cod War of 1972–1973. Research Series no. 29, Institute of International Studies, University of California, Berkeley.

Hatcher, A., and S. Pascoe. 2002. Future Options for UK Fish Quota Management. Centre for the Economics and Management of Aquatic Resources, University of Portsmouth.

Heizer, S. 2000. "The Commercial Geoduck Fishery in British Columbia, Canada— An Operational Perspective of a Limited Entry Fishery with Individual Quotas." In Use of Property Rights in Fisheries Management: Workshop Presentations, ed. R. Shotton (fisheries technical paper 404/2, FAO).

Hersoug, B. 2002. *Unfinished Business*. Eburon.

Hersoug, B., P. Holm, and S. A. Rånes. 2000. "The Missing T: Path-Dependency within an Individual Vessel Quota System—The Case of Norwegian Cod Fisheries." In Use of Property Rights in Fisheries Management: Workshop Presentations, ed. R. Shotton (fisheries technical paper 404/2, FAO).

Hirshleifer, J. 1982. "Evolutionary Models in Economics and Law." In Evolutionary Models in Economics and Law, ed. R. Zerbe (*Research in Law and Economics* 4).

Hjertonsson, K. 1973. *The New Law of the Sea*. Norstedt.

Hollick, A. 1977. "The Origins of 200-Mile Offshore Zones." *American Journal of International Law* 71: 494–500.

Hollick, A. 1981. *U.S. Foreign Policy and the Law of the Sea*. Princeton University Press.

Huppert, D. D. 1988. Managing Alaska Groundfish: Current Problems and Management Alternatives. Report FRI-UW-8805, Fisheries Research Institute, University of Washington.

Huppert, D. D. 1991. "Managing the Groundfish Fisheries of Alaska: History and Prospects." *Reviews in Aquatic Sciences* 4: 339–373.

Iyambo, A. 2000. "Managing Fisheries with Rights in Namibia: A Minister's Perspective." In Use of Property Rights in Fisheries Management: Mini-Course Lectures and Core Conference Presentations, ed. R. Shotton (fisheries technical paper 404/1, FAO).

Jóhannesson, J. 1956. *Íslendinga saga*, volume 1. Almenna Bókafélagid.

Kidd, D. 2000. "A Minister's Perspective on Managing New Zealand Fisheries." In Use of Property Rights in Fisheries Management: Mini-Course Lectures and Core Conference Presentations, ed. R. Shotton (fisheries technical paper 404/1, FAO).

Klebnikov, P. 2000. *Godfather of the Kremlin*. Harcourt.

Leblanc, J. 2000. "United States' Fishery Cooperatives: Rationalizing Fisheries through Privately Negotiated Contracts." In Use of Property Rights in Fisheries Management: Workshop Presentations, ed. R. Shotton (fisheries technical paper 404/2, FAO).

MacGowan, D., ed. 2001. *The Stonemason: Donald Macleod's Chronicle of Scotland's Highland Clearances*. Praeger.

Mackie, J. D. 1964. *A History of Scotland*. Penguin.

Matulich, S. C., R. C. Mittelhammer, and C. Reberte. 1996. "Toward a More Complete Model of Individual Transferable Fishing Quotas: Implications of Incorporating the Processing Sector." *Journal of Environmental Economics and Management* 31: 112–128.

Matulich, S. C., M. Sever, and F. Inaba. 2001. "Fishery Cooperatives as an Alternative to ITQs: Implications of the American Fisheries Act." *Marine Resource Economics* 16: 1–16.

Mayekiso, M., R. Tilney, and J. de Swardt. 2000. "Fishing Rights in South Africa." In Use of Property Rights in Fisheries Management: Mini-Course Lectures and Core Conference Presentations, ed. R. Shotton (fisheries technical paper 404/1, FAO).

McCay, B., and Brandt, S. 2002. "Changes in Fleet Capacity and Ownership of Harvesting Rights in the U. S. Surf Clam and Ocean Quahog Fishery." In Case Studies of the Effects of Introduction of Transferable Property Rights on Fleet Capacity and Concentration of Ownership in Marine Fisheries. FAO.

Mikalsen, K. H., and S. Jentoft. 2003. "Limits to Participation? On the History, Structure and Reform of the Norwegian Fisheries Management." *Marine Policy* 27: 397–407.

Miles, E. L. 1998. *Global Ocean Politics: The Decision Process at the Third United Nations Conference on the Law of the Sea 1973–1982*. Martinus Nijhoff.

Moloney, D. G., and P. H. Pearse. 1979. "Quantitative Rights as an Instrument for Regulating Commercial Fisheries." *Journal of the Fisheries Research Board of Canada* 36: 859–867.

NRC (National Research Council). 1999. *Sharing the Fish: Toward a National Policy in Individual Fishing Quotas*. National Academy Press.

Neher, P. A., R. Arnason, and N. Mollett, eds. 1989. *Rights Based Fishing*. Kluwer.

Newell, R. G., J. N. Sanchirico, and S. Kerr. 2003. Fishing Quota Markets. Discussion paper, Resources for the Future.

Nielander, W. J., and M. S. Sullivan. 2000. "ITQs—New Zealand and United States: Allocation Formula and Legal Challenges." In Use of Property Rights in Fisheries Management: Workshop Presentations, ed. R. Shotton (fisheries technical paper 404/2, FAO).

NMFS/NOAA (National Marine Fisheries Service and National Oceanic and Atmospheric Administration). 2002. Final Environmental Impact Statement for American Fisheries Act Amendments 61/61/13/8. US Department of Commerce.

Nordquist, M. H., ed. 1985–. United Nations Convention on the Law of the Sea 1982: A Commentary. Martinus Nijhoff.

Ostrom, E. 1990. Governing the Commons. Cambridge University Press.

Peña-Torres, J. 1995. Chilean Fisheries Regulation: A Historical Perspective. Working paper 135, Departemento de economía, Universidad de Chile.

Peña-Torres, J. 1997. "The Political Economy of Fishing Regulation: The Case of Chile." Marine Resource Economics 12: 253–280.

Peña-Torres, J. 2002. "Debates sobre Cuotas Individuales Transferibles: ¿'Privatizando' el mar? ¿Subsidios? o ¿Muerte anunciada de la pesca extractiva en Chile?" Revista de Estudios Públicos 86: 183–222.

Plé, J.-P. 2000. "Responding to Non-Member Fishing in the Atlantic: The ICCAT and NAFO Experiences." In Law of the Sea, ed. H. Scheiber. Martinus Nijhoff.

Pohl, R. G. 1985. "The Exclusive Economic Zone in the Light of Negotiations of the Third UNCLOS." In The Exclusive Economic Zone, ed. F. Vicuña. Westview.

Porter, R. M. 1996. Structural and Market Consequences of Harvest Quotas in Canada's Pacific Halibut Fishery. Presented at conference on Fisheries Population Dynamics and Management, University of Washington.

Rice, J. 2003. The British Columbia Rockfish Trawl Fishery. Paper given at FAO workshop on unsustainable fisheries, Mauritius.

Richards, E. 1982. A History of the Highland Clearances. Croom Helm.

Richards, E. 1999. Patrick Sellar and the Highland Clearances. Polygon.

Robinson, W. L. 1986. "Individual Transferable Quotas in the Australian Southern Bluefin Tuna Fishery." In Fishery Access Control Programs Worldwide, ed. N. Mollett. Sea Grant Program, University of Alaska, Fairbanks.

Rogers, P. P. 2000. "Towards a Better Future Fisheries Management: Rights-Based Fisheres Management in Western Australia." In Use of Property Rights in Fisheries

Management: Mini-Course Lectures and Core Conference Presentations, ed. R. Shotton (fisheries technical paper 404/1, FAO).

Rousseau, J.-J. 1755. *A Discourse on Inequality*. Penguin, 1984.

Ruddle, K. 1984. "The Continuity of Traditional Management Practices: The Case of Japanese Coastal Fisheries." In *The Traditional Management of Coastal Systems in Asia and the Pacific*, ed. K. Ruddle and R. Johannes. UNESCO.

Ruddle, K., and R. E. Johannes. 1984. *The Traditional Management of Coastal Systems in Asia and the Pacific*. UNESCO.

Schelling, T. C. 1978. *Micromotives and Macrobehavior*. Norton.

Scheiber, H. N., ed. 2000. *Law of the Sea: The Common Heritage and Emerging Challenges*. Martinus Nijhoff.

Schrank, W. E., R. Arnason, and R. Hannesson, eds. 2003. *The Cost of Fisheries Management*. Ashgate.

Scrutton, T. E. 1887. *Commons and Common Fields*. Burt Franklin, 1970.

Sebenius, J. K. 1984. *Negotiating the Law of the Sea*. Harvard University Press.

Shotton, R., ed. 2000a. Use of Property Rights in Fisheries Management: Mini-Course Lectures and Core Conference Presentations. Fisheries technical paper 404/1, FAO.

Shotton, R., ed. 2000b. Use of Property Rights in Fisheries Management: Workshop Presentations. Fisheries technical paper 404/2, FAO.

Smith, P. J. 2000. "How 'Privatization' Can Result in More Government: The Alaska Halibut and Sablefish Experience." In Use of Property Rights in Fisheries Management: Mini-Course Lectures and Core Conference Presentations, ed. R. Shotton (fisheries technical paper 404/1, FAO).

Smith, T. 1994. *Scaling Fisheries*. Cambridge University Press.

Steen, S. 1930. *Det norske folks liv og historie gjennom tidene*, volume 5. Aschehoug, Oslo.

Stevenson, G. G. 1991. *Common Property Economics*. Cambridge University Press.

Stollery, K. 1986. "A Short Run Model of Capital Stuffing in the Pacific Halibut Fishery." *Marine Resource Economics* 3: 137–153.

Sturgess, N., and F. Meany, eds. 1982. *Policy and Practice in Fisheries Management*. Australian Government Publishing Service.

Sullivan, J. M. 2000. Harvesting Cooperatives and U. S. Antitrust Law: Recent Developments and Implications. Paper given at biennial meeting of International Institute for Fisheries Economics and Trade, Corvallis, Oregon.

Tsamenyi, M., and A. McIlgorm. 2000. "Enhancing Fisheries Rights through Legis-lation—Australia's Experience." In Use of Property Rights in Fisheries Management: Workshop Presentations, ed. R. Shotton (fisheries technical paper 404/2, FAO).

Turris, B. R. 2000. "A Comparison of British Columbia's ITQ Fisheries for Ground-fish Trawl and Sablefish: Similar Results from Programs with Differing Objectives, Designs and Processes." In Use of Property Rights in Fisheries Management: Mini-Course Lectures and Core Conference Presentations, ed. R. Shotton (fisheries technical paper 404/1, FAO).

Ulltang, Ø. 1980. "Factors Affecting the Reaction of Pelagic Fish Stocks to Exploitation and Requiring a New Approach to Assessment and Management." *Conseil International pour l'Exploration de la Mer, Rapports et Procès Verbaux des Réunions* 177: 489–504.

Valatin, G. 2000. "Development of Property Rights-Based Fisheries Management in the United Kingdom and the Netherlands." In Use of Property Rights in Fisheries Management: Workshop Presentations, ed. R. Shotton (fisheries technical paper 404/2, FAO).

Vicuña, F. O., ed. 1984. *The Exclusive Economic Zone.* Westview.

Wang, S. 1995. "The Surf Clam ITQ Management: An Evaluation." *Marine Resource Economics* 10: 93–98.

Weitzman, M. L. 2002. "Landing Fees vs Harvest Quotas with Uncertain Fish Stocks." *Journal of Environmental Economics and Management* 43: 325–338.

Weninger, Q. 1998. "Assessing Efficiency Gains from Individual Transferable Quotas: An Application to the Mid-Atlantic Surf Clam and Ocean Quahog Fishery." *American Journal of Agricultural Economics* 80: 750–764.

Wertheimer, A. C., and D. Swanson. 2000. "The Use of Individual Fishing Quotas in the United States' EEZ." In Use of Property Rights in Fisheries Management: Workshop Presentations, ed. R. Shotton (fisheries technical paper 404/2, FAO).

Wilen, J. 2002. "Property Rights and the Texture of Rents in Fisheries." Paper given at workshop on Evolving Property Rights in Marine Fisheries, Center for Free Market Environmentalism, Bozeman, Montana.

Yamamoto, T., and K. Short, eds. 1992. *International Perspectives on Fisheries Management with Special Emphasis on Community-Based Management Systems Developed in Japan.* National Federation of Fisheries Cooperative Associations in Association with Japan International Fisheries Research Society.

Index

Absentee ownership, 104, 139, 143
Access to fish, 133
 controls on, 43, 135
 free, 43, 48, 53, 135
 open, 43, 50, 53, 93, 95, 145, 149,
  157, 160, 161, 171, 172
 rights of, 36, 43, 64, 82, 83, 126, 129
Alaska crab fishery, 5, 158–160
Alaska halibut fishery, 5, 79, 108, 109,
  136, 140–145, 160, 164, 166
Alaska pollock, 5, 51, 66, 74, 82,
  147–158, 160, 165, 168, 171
American Fisheries Act, 6, 23, 153, 154,
  158
Anadromous species, 37
Anchoveta, 67, 68
Anchovy, 61, 95–97
Angelini group, 96, 97
Anti-Reflagging Act, 150
Antitrust law, 152, 153
Arrow squid, 93

Blue whiting, 42
Buy-back, 89
Buyout, 57, 149
Bycatch, 149, 153, 154

Capacity of fishing boats and fleets, 41,
  58–62, 70, 100–102, 106, 115–118,
  137, 159
Capelin, 51, 61, 71, 118, 121

Capital stuffing, 62, 70
Capitalism, 8–11
Catch history, 58, 59, 76, 96, 117–119,
  138–140, 146, 152, 163, 166
Catcher boats, 148, 150, 154–156
Chicago boys, 85, 95
Clearances, 3, 15, 16, 19–24, 27
Coastal states, 34–41, 53, 69, 73–75,
  150, 169
Cod, 51, 75, 77, 103, 109, 116, 147, 169
 Barents Sea (Arcto-Norwegian), 51,
  103, 105
 Icelandic, 5, 51, 67, 68, 114–123
 Northern, 110
Collective goods, 66
Common Fisheries Policy, 105
Common heritage, 35
Common land, 14
Common property, 3, 10, 24, 43, 53,
  54, 65, 82, 129, 131
Common resource, 81, 82, 104, 127,
  178
Commons, 1, 15, 16, 23, 24, 43, 44, 82
Commons Preservation Society, 18
Community-based fisheries
  management, 64
Community-based rights systems, 83
Community development quotas, 82,
  149, 154, 159
Concessions, 62, 63, 81, 99–107
Consensual approach, 36, 38

Conservation, 31, 56, 57, 61, 63, 91, 93, 120
Continental shelf, 31–37, 113, 116, 125, 141
Cooperatives, 5, 6, 25, 64, 69, 146, 151–158
Cost recovery, 90, 142
Crab, 5, 136, 158, 159, 165
Crofter Act, 22

Decommissioning, 100
Deep-sea fisheries, 98
Deep-sea fishing, 88, 91
Deep-sea mining, 39
Discarding, 58, 158
Distant water fishing, 36, 37, 40, 41, 150
Dolphins, 171
Donut Hole, 42, 51

Economic theory of property rights, 24–27, 74, 146
Effort control, 61, 62, 138
Effort quotas, 76, 119, 122
El Niño, 61, 96, 98
Enclosure(s), 3, 14–19, 23, 24, 27, 35, 39
Enterprise quotas, 98
Environmental advocacy groups, 136, 161
Environmental Defense Fund, 150
Environmentalists, 65–67, 91, 150, 162, 168, 171, 178
European Union, 169
Exclusive economic zone, 3, 23, 32, 38–41, 43, 51–56, 64, 72, 87, 147
Exclusive fisheries zone, 38
Extinction of fish stocks, 44, 47, 65

Factory trawlers, 148, 153–156
Fish-dependent communities, 165
Fisheries, jurisdiction over, 37
Fisheries management councils, 73, 135, 136, 160, 174

Fishery Conservation and Management Act, 38, 135, 137
Fishing communities, 134
Fishing effort, 55, 56, 61, 70, 116, 138, 147, 166, 172
Fishing mortality, 121
Fishing power, 138
Flag of convenience, 26
Four-mile limit, 113
Freedom of enterprise, 135
Freedom of navigation, 3, 39, 41
Free trade, 170, 176

Geographically disadvantaged countries, 40, 41
Great Depression, 11, 86
Greenpeace, 91, 150
Gulf Fisheries Management Council, 150

Habitats, of fish, 25, 162
Halibut, 5, 58, 79, 108–112, 136, 140, 152, 160, 161, 164–166, 174
Hanseatic League, 25
Herring, 25, 26, 47, 61, 75, 98, 99, 114–118
  Atlanto-Scandian, 42, 71
  North Sea, 71, 98
High seas, 25, 41–43, 51–54, 69, 75
High Seas Catchers' Cooperative, 154
Highgrading, 60, 158
Highly migratory species (stocks), 37, 40, 41, 51, 52
Hoki, 93
Horse mackerel, 96, 98

Import restrictions, 170
Industrial Revolution, 6, 9, 20, 66
Inshore-offshore allocation, 148, 152–154
Inshore processing, 155, 156
International Court of Justice, 26, 29, 113–116

Law of the sea, 1, 2, 26, 29–42, 72, 75, 113, 116
Law of the Sea Convention, 35–39, 72, 73
License limitation, 64, 69, 70
Licenses for fishing, 58, 60–62, 69–70, 75, 79, 80, 95, 109, 164, 172, 178
Loophole, 51, 52

Mackerel, 105, 106
Magnuson Act, 142, 150, 151
Magnuson-Stevens Act, 151, 168
Management costs, 58, 59, 71, 72, 92, 109, 110
Marine protected areas, 162, 168
Marine Stewardship Council, 93
Maximum sustainable yield, 46, 48, 91
Merino Law, 2, 97
Mid-Atlantic Council, 139
Mineral nodules, 35
Moratorium on ITQs, 147–153, 158, 159
Motherships, 148, 153–156

National Marine Fisheries Service, 73, 142
National Research Council, 151
Natural equilibrium, 45, 46
Neutrality, zone of, 32–34, 177
North Atlantic Fisheries Organization, 71
North Pacific Management Council, 142, 143, 152, 159, 160
Northeast Atlantic Fisheries Commission, 71
Norway lobster, 114

Ocean quahog, 5, 79, 136–140, 160, 164, 171
Oil, 3, 31, 34, 35, 60, 66, 125, 162
Open-access fisheries, 18, 44, 48–52, 82, 88, 173
Orange roughy, 52, 88–93, 98
Overcapacity, 41, 57, 69, 99, 106, 146, 149, 150, 169, 172

Overexploitation, 95
Overfishing, 25, 65, 80, 150, 155, 178
    biological, 44, 50
    economic, 44, 50

Pacific Halibut Commission, 141, 142
Pacific Management Council, 152
Patagonian toothfish, 98
Pollock, Atlantic, 147
Pollock Conservation Cooperative, 154, 155
Prawn, 98
Price of quotas, 93, 122, 123, 147
Processing of fish, 109, 110, 122, 133–137, 147–151, 156, 162, 165
Processing plants, shore-based, 148–153
Processing quotas, 159
Processors, 111, 138, 149, 150, 154–159, 162, 165, 172

Quota prices, 95, 122, 125
Quotas
    allocation of, 59, 63, 90, 106, 117, 131, 132, 140
    auctioning of, 59, 60, 97, 126, 133, 160, 172
    value of, 58, 60, 122, 133, 152, 172

Recreational fisheries, 18, 65
Recruitment, of fish, 45
Red snapper, 150, 151
Regional fisheries organizations, 41, 42
Rent, 14, 20, 30, 58, 74, 155–157, 172
    capture, 54, 59, 125, 126, 157, 160, 173, 178
    of quotas, 58, 93, 124
    maximization of, in fisheries, 50, 51, 55, 56
    tax on, 59, 60, 125, 126, 172
Rent seeking, 90
Resource rentals, 58, 59, 80, 89, 90
Resource rents, 59, 89
Rockfish, 108–110, 169

Sablefish, 76, 107–110, 136, 141–145
Salmon, 37, 68, 69, 75, 109, 138
Sardines, 61, 95–97
Sea lions, 66, 137, 162, 168, 171
Seabed Committee, UN, 35
Seabed mining, 34–36, 38, 41. *See also*
  Deep-sea mining
Semana Internacional, 32, 34
Sharks, 67
Shrimp, 83, 102, 121
Small-scale fisheries, 57, 64
Social welfare, 11
Socialism, 7–10, 14, 178
South Atlantic Fisheries Management
  Council, 146
Southern blue whiting, 93
Southern bluefin tuna, 57, 76
Spawning, 45
Steam trawler, 26
Stewardship, 60, 155
Stinting, 16, 72, 178
Stock abundance, 45
Stock assessment, 61, 80, 92
Straddling stocks, 41, 51, 52
Submarines, 39
Subsidies, 87, 88, 100, 170, 176
Subsistence fisheries, 64
Surf clam, 5, 57, 61, 76, 79, 81,
  136–141, 159, 160, 164, 166, 171
Surimi, 147
Surplus growth, 45–48, 127
Surplus value, 9
Sustainability, 110
Sustainable catch, 44–50
Sustainable fisheries, 147
Sustainable yield, 48

Taxes, 58–60, 126, 172
Territorial Use Rights in Fisheries, 63, 64
Territorial waters, 40
Three-mile limit, 26, 31, 34, 113, 114,
  135

Three-pie system, 5, 159, 162
Total allowable catch, 40, 58, 80, 126,
  154, 158
Total catch quota, 60, 90, 97, 99, 106,
  108, 110, 117, 120, 141, 147, 148,
  160, 161
Transferability, 57, 106, 107, 111, 117,
  140–143, 161, 162
Treaty of Waitangi, 90
Truman Proclamation, 31–34, 113, 177
Tuna, 37, 40, 42, 52, 75
Turbot, 52
TURFs, 63, 64
Twelve-mile limit, 114, 115
Two-hundred-mile limits (zones), 31,
  38–42, 69, 71–75, 87, 88, 103, 105,
  107, 114, 116, 124, 125, 141, 147,
  160, 168, 169. *See also* Exclusive
  economic zone
  origins of, 32–34, 177
  and property rights, 51–56
"Two-pie" system, 149, 154, 159, 162

Uncertainty, 167, 177
United Nations, 34–36
Use rights, 18, 54, 55, 60, 65, 69, 85,
  87, 90, 125, 136, 142, 146, 151, 169,
  177, 178
  gains and losses from, 163, 164
  initiative to, 72–75
  opposition to, 166
  political obstacles to, 167–172
  and property, 4, 77–81, 158
  territorial, 55, 56, 63, 64

Whaling, 32, 65, 91
Whiting, 5, 74, 151–153, 157
Wildlife, 18, 66, 168
Windfall gain, 58, 63
World War II, 8, 26, 31, 32, 34, 69, 88,
  116
Wreckfish, 74, 145, 146, 159, 171